新编职业英语系列

新编
服装英语

中等职业学校职业英语教材编写组

ENGLISH

高等教育出版社·北京

图书在版编目（CIP）数据

新编服装英语 / 中等职业学校职业英语教材编写组编. -- 北京：高等教育出版社，2020.7
ISBN 978-7-04-054294-3

Ⅰ. ①新… Ⅱ. ①中… Ⅲ. ①服装工业 - 英语 - 中等专业学校 - 教材 Ⅳ. ①TS941

中国版本图书馆 CIP 数据核字（2020）第 102859 号

策划编辑	李 森	责任编辑	李 森	封面设计	王 洋	版式设计 孙 伟
责任校对	原 平	责任印制	尤 静			

出版发行	高等教育出版社		网 址	http://www.hep.edu.cn
社 址	北京市西城区德外大街4号			http://www.hep.com.cn
邮政编码	100120		网上订购	http://www.hepmall.com.cn
印 刷	廊坊十环印刷有限公司			http://www.hepmall.com
开 本	787mm×1092mm 1/16			http://www.hepmall.cn
印 张	8.25			
字 数	230千字		版 次	2020年7月第1版
购书热线	010-58581118		印 次	2020年7月第1次印刷
咨询电话	400-810-0598		定 价	25.00元

本书如有缺页、倒页、脱页等质量问题，请到所购图书销售部门联系调换
版权所有 侵权必究
物 料 号 54294-00

前　言

英语是当今世界广泛使用的通用语言，是国际交流与合作的重要工具，是思想与文化的载体。学习英语对人的全面发展有积极的促进作用。中等职业学校英语教学需要贴近学生生活和未来岗位，帮助学生了解中外职场文化和中外优秀企业文化，培养学生的职业精神和工匠精神，提升学生的职业素养，为学生的终身学习和职业发展服务。因此，我们需要能够与各个专业充分结合的职业英语系列教材。本系列中的《新编服装英语》，是在教育部职业教育与成人教育司推荐的中等职业学校服装类专业英语教材基础上修订而成的，可供中等职业学校或高等职业院校服装类专业及相关专业学生使用，也可供服装行业岗位培训和自学者选用。

本教材以培养应用型人才为目标，结合服装类专业学生毕业后实际工作岗位技能的需求，向学生提供其未来工作岗位所需的服装行业的英语知识与技能训练。

本教材在编写前对中外服装设计和制造的工艺流程进行了大量调研，在编写中改变了以往教材按照学科体系进行编写的方法，本着学习与工作一体化的原则，根据服装行业的主要特性及工艺流程，在体例编排上尽量体现工作过程导向的思想。教材内容包括：设计灵感的产生、设计图纸的完成、服装打版方法及其技巧、样衣制作、面料选择、缝制技巧、服装配饰、服装展示、市场营销、作品集的收录、服装职业前景的分析与预测等。教材内容不仅围绕工作过程展开，同时还涉及职业前景等相关知识，为学生进入该行业打好基础。

本教材具有如下特色：

1. 反映服装产业发展现状及其前景

教材在市场调研的基础上以实例的形式深入浅出地介绍了服装设计与制作、服装面料的选择与设计、手工与机器缝制技巧等服装领域各个环节的专业基础知识，并在此基础上提供了丰富的关于服装配饰、服装流行趋势与展示、销售策略等方面的最新学习资料。这些资料信息量大，生动有趣，富有鲜明的时代特色。

2. 职业特色鲜明

教材编写突出职业英语的特点，尽可能将语言活动设置在真实场景中。例如，在学习"设计灵感产生"这一单元时，其中一项教学内容是让学生根据所搜集到的素材介绍灵感来源，并结合所学专业，将灵感以草图形式表现出来。这样，学生不仅提高了语言的运用能力，还加深了对未来工作任务的了解，从而强化专业知识技能，满足未来工作需求。

3. 选材真实、内容实用

全文选材真实，语言地道，内容生动、实用，既能反应服装设计、制作、销售的工作过程，又能突出训练学生动手、动口能力。语言难度贴近学生的水平和需求。

教学建议：

1. 课时安排

通过教学实验，建议每个单元安排 4~6 课时：词汇及听力 1~2 课时，阅读及语言知识 2 课时，会话及活动 1~2 课时。

2. 单元教学建议

本教材共 11 个单元。每个单元包含：

第一部分：Vocabulary，这部分内容主要是介绍与本单元有关的专业英语词汇，以听力选择等形式呈现，旨在引导学生通过学习词汇了解本单元学习的主题。

第二部分：Listening，这部分内容先以完整对话的形式呈现，以降低学习的难度，再通过练习的形式帮助学生提高听力水平。

第三部分：Reading，这部分内容主要介绍与主题有关的服装领域的知识，例如，如何捕获设计灵感、如何以草图的形式展现设计灵感、服装打版的方法及技巧、样衣制作的方法、服装缝纫的技巧等，再通过阅读理解练习检测学生的阅读能力。

第四部分：Speaking，这部分内容是在学习前三部分内容的基础上围绕专业知识进行的小组讨论活动。学生可以借助词典、工具书和网络资源等完成学习任务，必要时也可以适量使用中文。在整个学习过程中，教师应及时给予指导。

第五部分：Language study，这部分内容是对语言知识的训练，旨在提高学生的实际运用水平。

第六部分：Acting it out，这部分内容以工作任务的形式出现，每个任务均为服装设计及制作的一个工作环节。学生完成任务时以动口、动手操作为主，教师给予学生充分的语言、技术指导，可以允许学生用适量的中文进行必要的交流。各单元工作任务环环相扣，前面的任务为后面的任务作铺垫，例如在教授完第一单元设计灵感的产生时，该单元的任务要求是让学生搜集并介绍设计灵感，而第二单元的学习任务就是以草图的形式表现自己的设计灵感，直至学完本教材时请学生将自己最终完成的成衣作品进行展示并制定营销策略等。

第七部分：Reading for more，这部分内容是对各单元学习内容进行补充，可作为选读内容。教师应鼓励学生自学或在教师指导下学习。

第八部分：Self-check，这部分内容是学生自我测试，以检测学习效果。

3. 评价方式

在使用本教材时，教师可以学生每次完成的工作任务及最终展示的作品、课堂学习态度等作为考核的内容。

本教材由北京电子科技职业学院陈开宇担任主编，北京信息科技大学杨韩钰担任副主编，参加编写的还有杭州市服装职业高级中学刘雪蓉和大连模特艺术学校刘敏。陈开宇承担了本教材全书的审稿工作，第二、三、五、六、十、十一单元的编写工作以及其他单元的部分编写工作，并对全书进行了统稿；杨韩钰承担了本教材全书的审稿工作和统稿工作；刘雪蓉承担了本教材第一、四、九单元的部分编写工作；刘敏承担了本教材第七、八单元的部分编写工作。北京电子科技职业学院王明杰、田方两位老师对教材的编写提纲提出了大量修改建议。北京石油化工学院王笃勤教授、北京第二外国语学院美籍外教 Robert Scott Searer 对书稿进行了审阅。本教材在编写过程中还得到了北京教育科学研究院刘海霞、杭州市职业教育与成人教育研究室林海燕、大连教育学院于红三位老师的大力支持，在此一并表示感谢。

限于编者的经验和水平，难免有疏漏和失误之处，敬请同行专家和读者批评指正。

编写编

2020 年 3 月

CONTENTS

Unit 1   A Design Inspiration ·· 1

Unit 2   Sketch Design ·· 11

Unit 3   Pattern Draping and Drafting ································ 20

Unit 4   Developing a Sample Garment ······························ 30

Unit 5   Suitable Clothing Fabric ·· 40

Unit 6   Sewing Techniques ·· 50

Unit 7   Fashion Accessories ··· 59

Unit 8   Fashion Show Planning ··· 69

Unit 9   Garment Marketing ·· 79

Unit 10   Portfolio ·· 89

Unit 11   Career Expectations ··· 98

Words & Expressions ··· 108

Text Translations ·· 117

Unit 1

A Design Inspiration

Vocabulary

1. Listen and choose the correct word or expression in the box for each picture, change the form if necessary.

Where can you find inspiration?

| theater | photography | graphic | exhibition |
| fabric swatch | trimming | magazine | color palette |

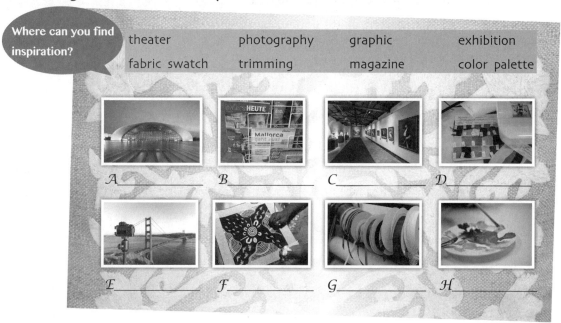

A_____ B_____ C_____ D_____

E_____ F_____ G_____ H_____

1

2. Use the correct word or expression from the box to fill in each blank.

1) A _____ is a building where plays are performed on a stage.

2) An _____ is a collection of pictures, sculptures or other things displayed in a public place.

3) The _____ on a piece of clothing are extra parts added for beauty or decoration.

4) A fabric _____ is a small piece of cloth used as an example of a larger piece.

5) _____ are drawings and pictures that are made using simple lines and sometimes strong colors.

Listening

Mary:	Wow! So cool! How wonderful your design is!
Tim:	Really? I am happy you like it.
Mary:	Where did you get the inspiration for it?
Tim:	I got it from a car.
Mary:	A car? Are you kidding, Tim?
Tim:	Ha ha. The beautiful and clear lines of a car gave me the inspiration.
Mary:	Yeah. You are so great! It's always difficult for me to find inspiration.
Tim:	Well, I think inspiration is everywhere. Once you find something new, you'd better take notice as soon as possible. Use your eyes and follow your heart, then you can find what you want.
Mary:	Yeah, you are very helpful. Thank you very much.

Unit 1 A Design Inspiration

1. **Listen to the dialogue and tick the right picture to answer the question.**
   Where did Tim get his inspiration?

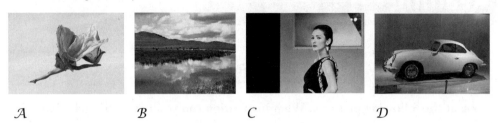

A   B   C   D

2. **Listen to the dialogue again and decide if the following sentences are True (T) or False (F).**

   T F
   ☐ ☐ 1) The beauty and the color of the car gave Tim the inspiration.
   ☐ ☐ 2) Mary didn't believe that the inspiration for Tim's design came from a car.
   ☐ ☐ 3) Both Mary and Tim can be inspired to design different styles.
   ☐ ☐ 4) Tim thinks everything can inspire people to create.
   ☐ ☐ 5) Mary gets some help from Tim.

3. **Listen to the dialogue again and complete the sentences with the expressions in the box.**

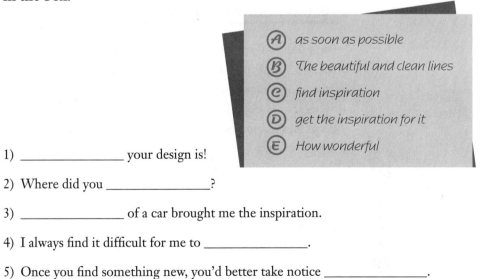

   Ⓐ as soon as possible
   Ⓑ The beautiful and clean lines
   Ⓒ find inspiration
   Ⓓ get the inspiration for it
   Ⓔ How wonderful

   1) _____ your design is!

   2) Where did you _____?

   3) _____ of a car brought me the inspiration.

   4) I always find it difficult for me to _____.

   5) Once you find something new, you'd better take notice _____.

3

Reading

Before You Read

Look at the following pictures. What inspiration can you get from each picture?

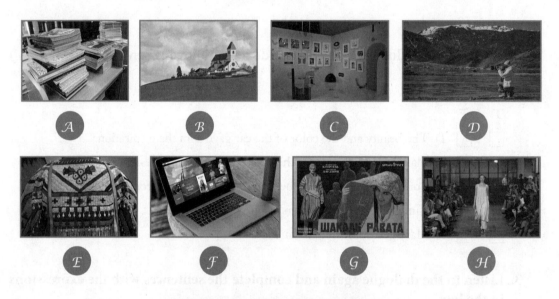

A B C D

E F G H

Sources of Design Inspiration

For a designer inspiration can be drawn from almost everywhere. Books and magazines, exhibitions, art shows, world happenings, theaters, music, dance, Internet, family photographs and world travel are all sources of design inspiration.

Books and magazines can often help us find information and show us photographs of different styles. Trade magazines are a good source of up-to-date design trends, catwalk news and fashion comment. They provide information on new developments in fabric technology.

Exhibitions can be inspiring. During the last 30 years there have been exhibitions of Native American, Mexican, Egyptian and French art. They have had a direct influence on fashion design.

Experiencing different cultures provides inspiration.

Graphics and photography can provide a rich source of

Unit 1　A Design Inspiration

inspiration both for design and illustration purposes.

　　Raw materials in themselves can be the main source of inspiration — fabric swatches, color palettes, trimmings, old wallpapers, antique fabrics.

　　A good designer knows how to read clues from the sources and produce designs that are fresh and unique.

Words & Expressions

antique /æn'tiːk/ *adj.* 古老的，古董的
catwalk /'kætwɔːk/ *n.* （时装表演时模特走的）伸展台
comment /'kɒment/ *n.* 评论
exhibition /ˌeksɪ'bɪʃən/ *n.* 展览会
fabric /'fæbrɪk/ *n.* 面料
graphics /'græfɪks/ *n.* 图样，图案
illustration /ˌɪlə'streɪʃən/ *n.* （书、杂志等中的）图表，插图
influence /'ɪnfluəns/ *n.* 影响；作用
inspiration /ˌɪnspə'reɪʃən/ *n.* 灵感
line /laɪn/ *n.* 细线；线条

palette /'pælɪt/ *n.* 调色板
photograph /'fəʊtəɡrɑːf/ *n.* 照片
photography /fə'tɒɡrəfɪ/ *n.* 照相术，摄影
source /sɔːs/ *n.* 资源
technology /tek'nɒlədʒɪ/ *n.* 技术
trend /trend/ *n.* 趋势
trimming /'trɪmɪŋ/ *n.* 装饰品，镶边饰物

fabric swatch 面料小样
fashion comment 时尚评论
raw material 原材料

Reading Comprehension

1. Read the passage again and match the sources with the information.

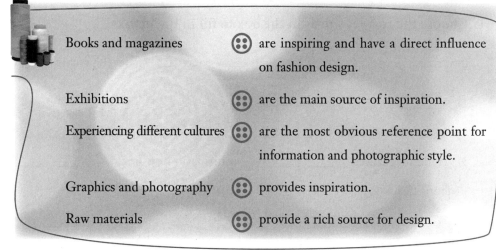

Books and magazines　　　are inspiring and have a direct influence on fashion design.

Exhibitions　　　are the main source of inspiration.

Experiencing different cultures　　　are the most obvious reference point for information and photographic style.

Graphics and photography　　　provides inspiration.

Raw materials　　　provide a rich source for design.

2. **Read the passage again and decide if the statements are True (T) or False (F).**

   T   F
   ☐   ☐   1) All the sources of inspiration are related to the designer's experience.
   ☐   ☐   2) Newspapers are good source on fabric technology.
   ☐   ☐   3) Designers must draw on many areas for inspiration.
   ☐   ☐   4) Designers have design block and they can't break it whatever they do.
   ☐   ☐   5) A good designer can derive new and fresh inspiration for their fashion design because they have their own sources.

Speaking

 1. Show some pictures to your partner and tell your partner what kinds of things inspire your designs.

2. How does the inspiration help you make a design?

Language study

1. **Choose the correct words in the box to fill in the blanks.**

inspiring	everywhere	provide	sources	developments	How

   1) _____ wonderful your design is!

   2) Books and magazines are _____ of design inspiration.

   3) Trade magazines _____ information on new _____ in fabric technology.

Unit 1 A Design Inspiration

4) Any exhibition can be _____ for a good designer.

5) Inspiration is _____.

2. **Complete the sentences with the words or expressions you have learned.**

   1) You are so great! But it is always _____ for me to find inspiration.

   2) _____ your eyes and _____ your heart, then you can _____ what you want.

   3) _____ you find _____ _____, you'd better take notice of them as _____ as possible.

   4) Magazines are a good _____ of _____ design trends.

   5) The big exhibition has had a direct _____ on fashion design.

3. **Choose the correct sentence in the box for each picture.**

   1) Do graphics and photography often help designers to derive a lot of information?
   2) If you feel inspiration, you should capture (捕获) it at once.
   3) Many sources of inspiration are related to a designer's personal experience.
   4) Every week John goes to see fashion shows to get inspiration.

Acting it out

1. Discuss with your classmates and decide what you want to design.
2. Tell your classmates where you get the inspiration.
3. Get some advice from your classmates on your design and revise your design.
4. Collect as much information from the other classmates and then complete the form.

Name	Fashion design	Inspiration comes from	Other sources of inspiration
Sarah	dress	a film	travel, books, fashion shows

Unit 1 A Design Inspiration

Reading for more

WHERE DOES INSPIRATION OF FASHION DESIGNERS COME FROM?

Fashion inspiration can be found in anything, even the smallest most ordinary things. Here are some sources for fashion inspiration:

1. Collecting inspiration in museums & art galleries

There are various types of art facts, objects and historical treasures in museums & art galleries. Those antiquities, furniture, sculptures, architectural models and drawings, and paintings can bring you a large source of primary research. It's also perfect for fashion designers or lovers to explore techniques from the past and add to their contemporary creations.

2. Being creative from music and films

Music can not only relax you but also bring you creative inspiration. Try to feel a piece of music and sketch your picture when you listen to it! Films are like music. You can find various fashion elements in a film. Using film as the starting point for primary research is something designers have been doing for years.

3. Going outside

A good fashion designer cannot finish design works only staying indoors. Going out to explore and learn from other cultures can provide you with a rich source of primary research material. Whether these places are travel destinations or just a flea market, as a fashion designer, discovering new and exciting possibilities around you can stretch your imagination further.

Inspiration doesn't always just happen at the snap of a finger. To be a good designer, please keep your eyes open and your mind active all the time!

Self-check

In this unit I've remembered the following words and expressions:

- ☐ inspiration
- ☐ exhibition
- ☐ source
- ☐ line
- ☐ photograph
- ☐ trend
- ☐ catwalk

- ☐ comment
- ☐ fabric
- ☐ technology
- ☐ influence
- ☐ graphics
- ☐ photography
- ☐ illustration

- [ ] palette
- [ ] trimming
- [ ] antique
- [ ] fashion comment
- [ ] raw material
- [ ] fabric swatch

I now understand and can use the following sentences:

_____

_____

10

Unit 2

Sketch Design

Vocabulary

1. Listen and choose the correct word or expression in the box for each picture, change the form if necessary.

What do the pictures mean?

| translucent paper | structure | proportion |
| surface decoration | fitting line | a layout pad |

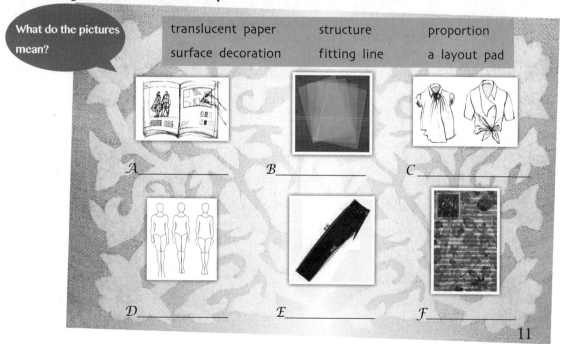

A_____ B_____ C_____

D_____ E_____ F_____

2. Read and match the two parts.

1) a layout pad   a. 净缝线
2) proportion   b. 外表装饰
3) fitting lines   c. 透明纸
4) surface decoration   d. 比例
5) translucent paper   e. 草图本

Listening

Li Lin: Mr Wang, will you please talk about your ideas on working as a fashion designer?

Mr Wang: Well, in my opinion, a designer's work is to select proper fabric and create fashionable designs.

Li Lin: Do you need to do some research before you begin a sketch design?

Mr Wang: Absolutely. The first step in creating a design is researching current fashion and predicting future trends.

Li Lin: What do you think of "design development is a challenge"?

Mr Wang: Well. A designer must not only be creative, he or she must also know what sells and how to get along with their coworkers.

Li Lin: So how do you define a good designer?

Mr Wang: I think a good designer must be creative, with smart insight into fashion. And, a designer should have a strong sense of the market, as well as a cooperative attitude.

Li Lin: Thanks for giving us so much advice.

Mr Wang: It's my pleasure.

Unit 2 Sketch Design

1. **Listen to the dialogue and tick the correct answers to the question.**

   What things / abilities does a designer have according to the dialogue?
   - [ ] do some research
   - [ ] know how to get along with companions
   - [ ] sell clothes successfully
   - [ ] be creative
   - [ ] have smart insight into fashion
   - [ ] have pattern-making skills
   - [ ] have a strong sense of the market

2. **Listen to the dialogue again and match the two parts to make a sentence.**

   1) A designer must know what
   2) A designer should have a strong
   3) The first step in creating a design is
   4) Will you please talk about
   5) A designer's work is to select

   a. your ideas on working as a fashion designer?
   b. proper fabric and create fashionable designs.
   c. sells and how to get along with their coworkers.
   d. sense of the market, as well as a cooperative attitude.
   e. researching current fashion and making predictions of future trends.

3. **Listen to the dialogue again and complete the sentences with the expressions in the box.**

   - Ⓐ with smart insight
   - Ⓑ giving us so much advice
   - Ⓒ In my opinion
   - Ⓓ do some research
   - Ⓔ have a strong sense of

   1) _____, a designer's work is to select proper fabric and create fashionable designs.

   2) Do you need to _____ before you begin a sketch design?

   3) A good designer must be creative, _____ into fashion.

4) A designer should _____ the market.

5) Thank you for _____.

Reading

Before You Read

Look at the following pictures. Discuss the question with your partner.
What do you think is the correct order of the pictures?

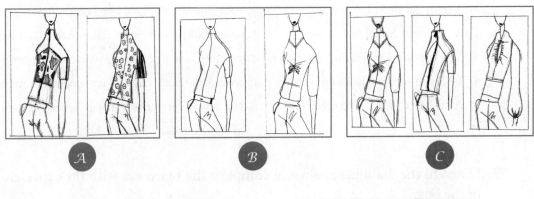

A   B   C

D

E

Unit 2 Sketch Design

Learn to Draw Your Ideas

As a designer, it is important to know how to develop your ideas on paper. The steps taken in drawing your ideas not only expresses your thoughts, but also helps you create new ideas by constant experimentation.

Here is a popular way of designing:

First, using a layout pad or translucent paper, trace your design ideas over the figure and vary the structure. During this period it is more important to concentrate on developing design ideas rather than on drawing the figure correctly.

In order to choose the best ideas, for the second step, you need to be clear about the direction and take on new ideas. Generally speaking, good ideas may take some time to hatch because the new idea itself is often challenging and initially looks "wrong" in some way. But don't give up. Keep your sense of judgment, but be optimistic.

When you are satisfied with the results, then you can complete the process, using form and shape to play with proportion, style or fitting lines so that a working structure is created.

Finally, don't forget to consider adding the surface decoration, detail and finish. They are also important for the best results of your experimentation.

During the whole process, you need to consider the selected fabrics, trim and color palette.

Words

decoration /ˌdekəˈreɪʃən/ n. 装饰；装饰品
fitting /ˈfɪtɪŋ/ adj. 合身的
judgment /ˈdʒʌdʒmənt/ n. 判断
hatch /hætʃ/ v. 策划；孵化
initially /ɪˈnɪʃəli/ adv. 开始；最初
layout /ˈleɪaʊt/ n. 布局；设计

optimistic /ˌɒptɪˈmɪstɪk/ adj. 乐观的
process /ˈprəʊses/ n. 过程；进程
proportion /prəˈpɔːʃən/ n. 比例；协调
surface /ˈsɜːfɪs/ n. 表面
structure /ˈstrʌktʃə/ n. 结构；构造
translucent /trænzˈluːsnt/ adj. 半透明的

Reading Comprehension

1. **Read the passage again and number the sentences according to the passage.**

   What are the steps of drawing ideas?
   - [ ] Complete the process, using form and shape to play with proportion, style or fitting lines.
   - [ ] Trace your design ideas over the figure and vary the structure.
   - [ ] Add the surface decoration, detail and finish.
   - [ ] Don't forget to consider the selected fabrics, trim and color palette.
   - [ ] Be clear about the direction and take on new ideas.

2. **Read the passage again and complete the steps of drawing ideas.**

   Trace _____ —> Be clear _____ —> _____ —>

   _____ —> _____

Speaking

1. Tell your partner what abilities of working as a designer he / she has.
2. Introduce your last sketch finished to your partner.
3. Tell your partner what research you had done before you finished the sketch.

Unit 2 Sketch Design

Language study

1. **Choose the correct words or expressions in the box to fill in the blanks.**

select	think of	get along with	working as	as well as

   1) What do you _____ "design development is a challenge"?

   2) A designer should have a strong sense of the market, _____ a cooperative attitude.

   3) A designer's work is to _____ proper fabric and create good designs.

   4) Will you please talk about your ideas on _____ a fashion designer?

   5) A designer should _____ their coworkers.

2. **Complete the sentences with the words or expressions you have learned.**

   1) It is more important to _____ on developing design ideas than on drawing the figure correctly.

   2) Choose the best ideas, you need to be _____ about the direction and _____ new ideas.

   3) _____ as a designer, _____ is important to know how to _____ your ideas on paper.

   4) When you are _____ with the results, you can _____ the process.

   5) Don't forget to _____ adding the surface decoration, detail and finish.

3. **Put the words and expressions in the correct order to make sentences.**

   1) a popular, Here, of designing, way, is

   _____.

   2) to hatch, ideas, some time, good, take, It may

   _____.

3) to play with, style or fitting lines, the process, and shape, proportion, Complete, using form

   _____.

4) important, how to, It is, develop your ideas, to know, on paper

   _____.

5) define, how, a good designer, So, you, do

   _____?

Acting it out

1. Show your sketch you've finished last time to your partner.

2. Talk with your partner about the sketch from different aspects, such as the structure, shape, proportion, trim and fabric and so on.

3. Ask your partner for some advice.

4. Write down their advice.

5. Revise your sketch until you are satisfied with it.

Unit 2 Sketch Design

Reading for more

HOW IS A DESIGN EVALUATED?

Most good designers need to follow the same routes to finish their designs: design brief, investigation, ideas for solutions, chosen solutions, realization and evaluation.

But how is a design evaluated? A particular dress may sell well in Beijing but not in Shanghai. It may sell well in spring, only at a certain time in the season in certain colors or in particular positions within a store.

Computerized allocation systems give information in time to the company about what styles are the fastest sellers so that the stock can be replaced efficiently.

Self-check

In this unit I've remembered the following words:

- ☐ translucent
- ☐ structure
- ☐ proportion
- ☐ surface
- ☐ decoration
- ☐ fitting
- ☐ layout
- ☐ hatch
- ☐ initially
- ☐ judgment
- ☐ optimistic
- ☐ process

I now understand and can use the following sentences:

_____

_____

Unit 3

Pattern Draping and Drafting

Vocabulary

1. Listen and choose the correct word or expression in the box for each part of the dress form.

   Do you know the names of each part of the dress form?

center front line	center back line
bustline	neckline
hipline	side seam
shoulder blade line	princess panel
princess line	waist line

   A_____ B_____
   C_____ D_____ E_____ F_____
   G_____ H_____ I_____ J_____

20

Unit 3 Pattern Draping and Drafting

2. **Read and match the two parts.**

   1) center front line   A. 侧缝线
   2) princess panel   B. 胸围线
   3) side seam   C. 公主片
   4) hipline   D. 前中心线
   5) bustline   E. 臀围线

Listening

(Mr Wang is giving the students a lesson about pattern making.)

Mr Wang: There are two methods of making a pattern: draping and drafting.

Student: What's the difference between the two methods?

Mr Wang: Generally speaking, the draping method is ideal for soft, flowing designs. While the drafting method is necessary for areas like sleeves and pant legs which would be difficult to drape on a stand.

Student: Do you mean that before we make patterns, we need to decide which method we should choose?

Mr Wang: Not really. As a pattern maker you can combine the two methods. It depends on the dictates of the design, as well as on your technical training.

1. **Listen to the dialogue and tick the right picture to answer the question.**
   Which picture shows the topic of the dialogue?

A   B   C   D

2. **Listen to the dialogue again and decide if the following sentences are True (T) or False (F).**

   T F
   - ☐ ☐ 1) The draping method is ideal for soft, flowing designs.
   - ☐ ☐ 2) The draping method is used for areas like sleeves and pant legs.
   - ☐ ☐ 3) The drafting method is a way of making a pattern.
   - ☐ ☐ 4) As a pattern maker, you can use both the methods.

3. **Listen to the dialogue again and complete the sentences with the expressions in the box.**

   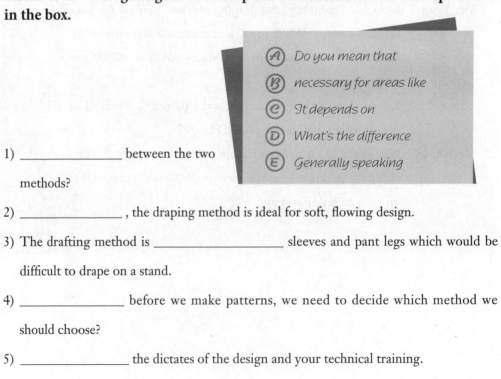

   Ⓐ Do you mean that
   Ⓑ necessary for areas like
   Ⓒ It depends on
   Ⓓ What's the difference
   Ⓔ Generally speaking

   1) _____ between the two methods?

   2) _____ , the draping method is ideal for soft, flowing design.

   3) The drafting method is _____ sleeves and pant legs which would be difficult to drape on a stand.

   4) _____ before we make patterns, we need to decide which method we should choose?

   5) _____ the dictates of the design and your technical training.

22

Unit 3 Pattern Draping and Drafting

Reading

Before You Read

Look at the following pictures. Discuss the question with your partner.
In what way were the patterns made?

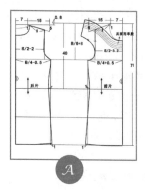
A

B

Pattern Draping and Drafting

Pattern making is an important step in the design process. The pattern maker can use either of the two methods for making patterns: draping or drafting.

The draping method needs to work with muslin on a dress form. The pattern maker cuts and shapes the fabric to a dress form so that the designer can see the proportions and lines of the design exactly. If necessary, the design can be altered on the form.

The pattern maker works on a dress form. He must carefully mark everything: center front, shoulder line, seams, armholes, buttonholes and so on. Draping techniques work best with jersey fabrics and generous amounts of soft and flowing materials. It is also used to work fabric on a bias.

Pattern drafting is a way of pattern making that depends on a series of form or figure measurements to complete the paper pattern. Drafting is blocking out on pattern paper a set of prescribed measurements for each piece. These patterns, after being tested for accuracy, become "blocks" or "slopers" that can be changed or adapted to each new style by moving darts and seams.

The drafting method is necessary for areas like sleeves and pant legs, which would be difficult to drape on a figure. Tailored garments are most successfully developed from flat pattern drafting.

Most pattern makers combine the two methods, basic shapes, such as bodices, sleeves, pants, or skirts, are draped or drafted. Choosing which method sometimes depends on the dictates of the design, as well as on their technical training and preference.

Words

armhole /ˈɑːmhəʊl/ n. 袖孔
bias /ˈbaɪəs/ n. 斜边；斜纹
bodice /ˈbɒdɪs/ n. 连衣裙的上部；紧身胸衣
buttonhole /ˈbʌtnhəʊl/ n. 纽扣孔
combine /kəmˈbaɪn/ v. 结合，组合
dart /dɑːt/ n. 缝褶
drape /dreɪp/ v. 把（织物）披在⋯上
draft /drɑːft/ v. 起草，草拟
flow /fləʊ/ v. 垂；飘拂
garment /ˈɡɑːmənt/ n. 服装
jersey /ˈdʒɜːzɪ/ n. 平针织物；紧身套衫
mark /mɑːk/ v. （在⋯上）做记号
measurement /ˈmeʒəmənt/ n. 尺寸；大小

muslin /ˈmʌzlɪn/ n. 平纹细布
prescribed /prɪˈskraɪbd/ adj. 规定的；指定的
procedure /prəˈsiːdʒə/ n. 步骤；手续
preference /ˈprefərəns/ n. 偏爱；钟爱
seam /siːm/ n. 线缝；缝口
shape /ʃeɪp/ v. 使成形；塑造
shoulder /ˈʃəʊldə/ n. 肩；肩膀
sleeve /sliːv/ n. 袖子
slope /sləʊp/ n. 斜面
stand /stænd/ n. 台；架
tailored /ˈteɪləd/ adj. （衣服）合身的；专为⋯而做的
technical /ˈteknɪkl/ adj. 技术的；工艺的

Reading Comprehension

1. Read the passage again and decide if the statements are True (T) or False (F).

T F

☐ ☐ 1) Draping and drafting are two methods for making patterns.

☐ ☐ 2) The drafting method needs to be used on a dress form.

☐ ☐ 3) Drafting techniques work best with soft materials and is necessary for areas like sleeves.

☐ ☐ 4) Pattern drafting is a way of pattern making that depends on some figure measurements.

Unit 3 Pattern Draping and Drafting

☐ ☐ 5) When the drafting method is used to make a pattern, the design can be altered on the form.

☐ ☐ 6) Pattern makers can combine the two methods.

2. Read the passage again and complete the table about draping and drafting methods.

Draping method	Drafting method
a. work with muslin	a.
b. on a dress form	b.
c.	c.
...	...

Speaking

1. Look at the dress form and tell your partner the names of each part on the dress form.

2. Here are some swatches for you and your partner. Please have a talk to study what methods you will choose for the fabrics to make patterns.

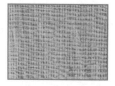  

25

Language study

1. Choose the correct words in the box to fill in the blanks.

| altered | while | combine | difference | depends |

1) What's the _____ between the two methods?

2) Pattern drafting _____ on a series of form or figure measurements.

3) As a pattern maker, you can _____ the two methods.

4) The draping method is ideal for soft designs _____ the drafting method is better for the parts difficult to drape on a stand.

5) The design can be _____ on the form when the draping method is used for pattern making.

2. Complete the sentences with the words or expressions you have learned.

There are two methods of _____ a pattern: draping or drafting. Generally speaking, the _____ method is ideal for soft, _____ designs. While the drafting method is necessary for areas like sleeves and pant legs which _____ difficult to drape on a stand. As a pattern maker, you can _____ the two methods. It _____ on your technical training.

3. Put the words and expressions in the correct order to make sentences.

1) the designer, Is, satisfied, the look and fit, with

   _____?

2) materials, Draping techniques, with, work best, soft and flowing

   _____.

3) form or figure measurements, Pattern drafting, a series of, to complete the paper

26

Unit 3 Pattern Draping and Drafting

pattern, depends on

_____.

4) is necessary for areas like, which would be difficult, The drafting method, sleeves and pant legs, to drape on a figure

_____.

Acting it out

1. Practice making a pattern according to your sketch finished last time. Ask your teacher or your partner for help when you have problems.

2. Write down the problems and the solutions you have made while making patterns.

3. Talk about your patterns with your partner and ask for some advice.

4. Revise your patterns.

Problems	Solutions

Reading for more

MATERIALS FOR DRAPING AND DRAFTING

Muslin An inexpensive fabric is used to drape garments made of woven goods. Soft muslin will simulate the draping quality of natural or synthetic silk (化纤绸), lingerie (内衣) fabric, and fine cottons. Medium-weight muslin will simulate the draping quality of wool or medium-weight cottons. Coarse (粗的) muslin will simulate the draping quality of heavy-weight wools and cottons. Also, canvas (帆布) muslin will simulate the draping qualities of such heavy-weight fabrics as denim (牛仔布), fur, or imitation fur (仿皮绒).

Pattern drafting paper Strong, white drafting paper, with 1-inch (2.5 cm) grids of pattern dots, of a good quality and thickness available in rolls of various widths.

Self-check

In this unit I've remembered the following words:

- ☐ shoulder
- ☐ drape
- ☐ draft
- ☐ flow
- ☐ sleeve
- ☐ stand
- ☐ combine
- ☐ technical
- ☐ procedure
- ☐ muslin
- ☐ shape
- ☐ mark
- ☐ seam

- ☐ armhole
- ☐ buttonhole
- ☐ jersey
- ☐ bias
- ☐ measurement
- ☐ prescribed
- ☐ slope
- ☐ dart
- ☐ tailored
- ☐ garment
- ☐ bodice
- ☐ preference

Unit 3   Pattern Draping and Drafting

I now understand and can use the following sentences:

_____

_____

Unit 4

Developing a Sample Garment

Vocabulary

1. Listen and choose the correct word or expression in the box for each picture, change the form if necessary.

What is necessary for making a sample garment?

| the finished pattern | design staff | a sample maker |
| sewing machine | fashion design | craft paper | pattern draping |

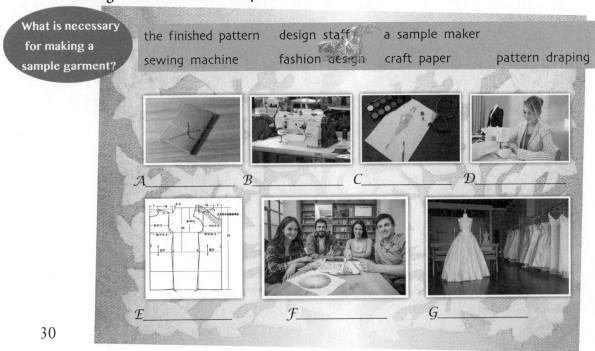

A_____ B_____ C_____ D_____

E_____ F_____ G_____

30

Unit 4 Developing a Sample Garment

2. Read and translate the Chinese terms into English.

1) 缝纫机   _____

2) 服装设计   _____

3) 设计人员   _____

4) 样衣工   _____

5) 牛皮纸   _____

Listening

Bill: Anna! Is the finished pattern OK?

Anna: Oh, I'm afraid not, Bill. There are some problems.

Bill: What's wrong? We have already altered the design many times.

Anna: I don't know. The pattern maker said he would discuss it with you in a while.

Bill: That's all right. Then, how about the fabrics?

Anna: We got them. Please take a look.

Bill: Oh! No, no. They are not what I want. They don't fit my design. I need some thinner fabrics and the colors should be brighter.

Anna: OK, we will try to find the proper ones as soon as possible.

Bill: Yeah. We'll need to hurry. We almost have no time to make the sample garment.

Anna: Bill, I have some bad news. The sample maker is ill. She has taken a three-day leave.

Bill: What? Oh, my god! That's too bad. Well, find another good maker. You know we must finish the sample garment by this Friday.

Anna: OK. We will do our best.

1. **Listen to the dialogue and tick the right pictures to answer the question.**
   What are mentioned in making a sample garment according to the conversation?

A a beautiful model

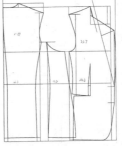

B the finished pattern

C suitable fabrics

D responsible design staff

E a good sample maker

F modern sewing machines

Unit 4　Developing a Sample Garment

2. Listen to the dialogue again and decide if the following statements are True (T) or False (F).

T　F
☐　☐　1) Bill is a fashion designer and Anna is his assistant.
☐　☐　2) Anna will discuss the problems about the design pattern with the pattern maker.
☐　☐　3) Bill is happy with the fabrics.
☐　☐　4) They have to find another sample maker because the former one is not good enough.
☐　☐　5) Time is limited for them to finish the sample garment.

3. Listen to the dialogue again and complete the sentences with the expressions in the box.

> Ⓐ find the proper ones
> Ⓑ what I want
> Ⓒ have already altered
> Ⓓ do our best
> Ⓔ thinner fabrics

1) We _____ the design many times.

2) They are not _____.

3) I need some _____ and the colors should be brighter.

4) We will try to _____ as soon as possible.

5) We will _____.

33

Reading

Before You Read

Look at the following pictures. Discuss the question with your partner.

What's the correct order to make a sample garment?

A

B

C

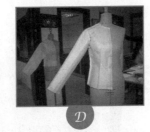
D

E

Making a Sample Garment

Making a sample garment is very important for developing new fashion designs and garments. Although it can be a tiresome and often boring process, it provides a finished pattern that is adapted to the models and can be used for countless finished garments.

A sample garment is made mainly through three steps, that is, paper pattern cutting, muslin sample checking and garment sewing.

Paper pattern cutting means to cut out the pattern which has been made of craft paper or blank paperboard according to the designs by the pattern maker. The size of the paper pattern is the same as a normal person.

Unit 4 Developing a Sample Garment

After the paper pattern has been cut out, it is laid out on the fabric — muslin, traced and cut out by the assistant designer or the sample cutter, that is muslin sample checking. This step puts the flat paper pattern into pattern draping to check the accuracy of the paper pattern. Some complex construction problems can be solved through the process. By constantly checking and altering, an accurate sample (the finished pattern) finally forms.

Garment sewing makes a sample garment with the fabrics, lining and trimmings really needed by the designers according to the finished pattern. A sample garment is made by a sample maker, the best of the factory sewers. He or she must not only know factory sewing methods but also know how to put an entire garment together. The design staff works closely with the sample maker to solve construction problems.

It is very important to see how the garment fits not only on a dress form but also on a model in motion. The designer can learn whether it is comfortable and looks good from the model's walking in the garment. Sometimes several samples have to be made to find the solution that fits the model best.

Words & Expressions

construction /kən'strʌkʃən/ n. 结构
flat /flæt/ adj. 平坦的
lining /'laɪnɪŋ/ n. 里料；里子
motion /'məʊʃən/ n. 移动；运动
sample /'sɑːmpl/ n. 样板；样品
solution /sə'luːʃən/ n. 解决办法
staff /stɑːf/ n. 全体成员
tiresome /'taɪəsəm/ adj. 令人厌倦的；索然无味的

a sample garment 一件成衣样品
blank paperboard 空白纸板
craft paper 牛皮纸
finished pattern 完成的样板
flat paper pattern 平面纸样
lay out 设计；安排

Reading Comprehension

1. Read the passage again and decide if the statements are True (T) or False (F).

T F

☐ ☐ 1) The writer thinks that making a sample garment is boring but very important for developing new fashion designs.

☐ ☐ 2) Muslin sample checking is one of the four steps to make a sample garment.

☐ ☐ 3) A sample garment is commonly made by the designer, because he / she knows the design very well.

☐ ☐ 4) Both the factory sewing methods and the skill of putting the entire garment together are necessary for a sample garment maker.

☐ ☐ 5) It is very important to see how the garments fit both on a dress form and on a model.

2. Read the passage again and fill out the form.
   How is a sample garment made?

The steps involved in making a sample garment.	Who does the work?	How to do the work?

Speaking

Before you make a sample garment, what preparations do you need to do? Talk about it with your partner.

36

Unit 4 Developing a Sample Garment

Language study

1. **Choose the correct expressions in the box to fill in the blanks.**

| has taken | the size of | have no time | something bad | the same as | laid out |

   1) We almost _____ to make a sample garment.
   2) I have to tell you _____. He is ill.
   3) After the paper pattern has been cut out, it is _____ on the fabric — muslin, traced and cut out.
   4) The pattern maker _____ a three-day leave.
   5) _____ the paper pattern is _____ a normal person.

2. **Complete the sentences with the words or expressions you have learned.**

   1) — They don't _____ my design. I need some thinner fabric.
      — We'll try to find some _____ ones as soon as possible.
   2) Put the flat paper pattern onto a pattern draping to check the _____ of the paper pattern.
   3) The design staff works _____ with the sample maker to solve _____ problems.
   4) The sample maker must know how to put the _____ garment together.
   5) Some construction problems can be _____ through the process.

3. **Choose the correct sentence in the box for each picture.**

   1) Has Xiao Wang been a sample maker for two years now?
   2) The muslin sample has been altered twice. Xiao Li is checking whether it is accurate now.
   3) The pattern cutter hasn't finished the paper pattern yet. She is very busy.

37

4) How many sample garments have they produced since 2020?

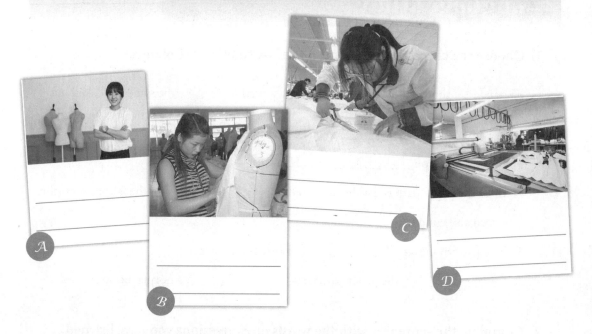

A

B

C

D

Acting it out

1. Lay out the garment on the fabric — muslin according to the pattern you've finished after learning Unit 3.

2. Show all the sample garments made by your classmates in the class. Give comments on them.

3. Take notes on the good points and the problems you find while enjoying the garment show.

Good points	Problems

Unit 4 Developing a Sample Garment

Reading for more

THE JOBS OF GARMENT PRODUCT DEVELOPMENT

1. Technical Designer

The technical designer translates the vision of the designer into reality. It's their job to make sure a particular garment can actually be manufactured efficiently. They must have creativity, excellent color and fashion sense, strong communication skills, and be knowledgeable about computers.

2. Pattern Maker

A pattern maker works closely with the designer to create master patterns for the desired design. Pattern makers must have good visualization skills, and be familiar with fabric and body construction. Experience is usually gained as an assistant to a pattern maker.

Self-check

In this unit I've remembered the following words and expressions:

- ☐ sample
- ☐ staff
- ☐ tiresome
- ☐ flat
- ☐ construction
- ☐ lining
- ☐ motion
- ☐ solution

- ☐ finished pattern
- ☐ a sample garment
- ☐ craft paper
- ☐ flat paper pattern
- ☐ blank paperboard
- ☐ lay out

I now understand and can use the following sentences:

_____

_____

39

Unit 5

Suitable Clothing Fabric

Vocabulary

1. **Listen and choose the correct word in the box for each picture, change the form if necessary.**

What fabrics are they?

| chiffon | cashmere | filoselle | jersey | twill |
| modal | silk | seersucker | denim | |

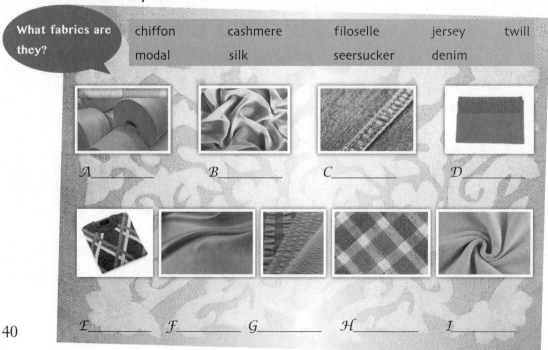

A_____ B_____ C_____ D_____

E_____ F_____ G_____ H_____ I_____

Unit 5 Suitable Clothing Fabric

2. **Look at the clothing labels below and write down the fabrics.**

M
72% Nylon
28% Spandex
Made in China
Style #23256

L
59% COTTON
3% MODAL
38% POLYESTER
Made in China

L
WASH COLD
25% CASHMERE
75% ACRYLIC
Made in China

S
HANG DRY
MACHINE WASH
95% COTTON
5% ELASTANE FIBER

Fabrics_____   _____   _____

_____   _____   _____

_____   _____   _____

Listening

(Mary's friend Jane is a good designer. Today Mary is going with Jane to a market that sells fabrics. She wants to learn how to choose fabrics.)

Mary: This is my first time to shop in a fabric market. I am not good at matching colors, fabrics and fashions. Can you give me some advice?

Jane: Sure. First you need to know the different kinds of fabrics and try to distinguish them.

Mary: I know a little about some common fabrics, like cotton, polyester, nylon, silk. But this fabric, I can't tell for sure, but it almost looks like a kind of seersucker to me.

Jane: Good. You have a good sense in distinguishing fabrics.

Mary: Is this kind of seersucker suitable for your designs?

> **Jane:** No, not really. For my pants sets, denim or twill is a good choice. But for the full dress and the evening gown, silk, brocade chiffon is much more suitable.
> **Mary:** But how do you recognize the quality of fabrics?
> **Jane:** Well. There are several ways to tell. The weave should be firm and uniform. The print color must be even. And, the dye color should be even and look fresh. Fabric should shed wrinkles after crushing.
> **Mary:** Great. I've learned so much from you today. Let's go to that shop and take a look.
> **Jane:** All right.

1. **Listen to the dialogue and tick the fabrics mentioned in it.**

cashmere	chiffon	denim	cotton	filoselle	jersey
modal	seersucker	nylon	twill	silk	polyester

2. **Listen to the dialogue again and tick the sentences mentioned in it.**

   - [ ] 1) Weave should be uniform.
   - [ ] 2) Print colors should be even.
   - [ ] 3) Weave should be firm.
   - [ ] 4) Dye colors should be even and look fresh.
   - [ ] 5) No powdery dust should appear.
   - [ ] 6) Fabric should shed wrinkles after crushing.

3. **Listen to the dialogue again and complete the sentences with the expressions in the box.**

   Ⓐ firm and uniform
   Ⓑ much more suitable
   Ⓒ some common fabrics
   Ⓓ have a good sense
   Ⓔ tell for sure

   1) I know a little about _____.

Unit 5 Suitable Clothing Fabric

2) I can't _____, but it almost looks like a kind of seersucker.

3) You _____ in distinguishing fabrics.

4) For the full dress and the evening gown, silk, brocade chiffon is _____.

5) The weave should be _____.

Reading

Before You Read

Look at the following pictures. Discuss the question with your partner.

Can you tell what fabrics are used to make these clothes?

A

B

C

D

Various Fabrics

Fabric is the material used to make garments. As one of the basic elements, fabric plays an important part in the style of garments. And in some degree it decides the color and pattern directly.

Fabrics can be classified into cotton, linen, wool, silk, man-made fiber fabric, jersey, leather. Different fabrics have their own characteristics.

Cotton is often used as a warm-weather fabric because it is light and washable. Cotton is often used in the fashion dress, leisure wear, underwear and shirts though it is sometimes easy-frilling and easy-shrinking.

Linen conducts heat and absorbs moisture. It is used as a fabric in leisure wear, gabardine and summer dress.

Wool is used for autumn and winter because it is heavy, wearable, soft, elegant and warm. It is often used to make high-quality garments, such as a full dress or a business suit. But wool is rather hard to wash or clean.

Silk is light weight, soft and luxurious, colorful but fades easily. It is usually used to make high-quality garments, especially women's wear.

Man-made fiber fabrics are often used for low to medium priced garments. Though they have poor friction, they are popular because they are colorful, soft, durable and comfortable.

Leather is usually applied to women's garments and winter garments. It's heavy, warm and graceful. But leather is expensive and hard to store. It requires a lot of care.

Fabrics themselves often inspire garment design. Some say, "Many a dress of mine was born of the fabric alone."

Words

brocade /brəʊˈkeɪd/ n. 锦缎
chiffon /ˈʃɪfɒn/ n. 雪纺绸；薄绸
conduct /kənˈdʌkt/ v. 传导（热或电）
cotton /ˈkɒtn/ n. 棉布
crush /krʌʃ/ v. 压坏；压伤
denim /ˈdenɪm/ n. 斜纹粗棉布
dye /daɪ/ n. 染料；染色
elegant /ˈelɪgənt/ adj. 优雅的；文雅的
even /ˈiːvn/ adj. 均匀的
fade /feɪd/ v. 褪色；凋落
fiber /ˈfaɪbə/ n. （动植物的）纤维
firm /fɜːm/ adj. 牢固的，稳固的
friction /ˈfrɪkʃən/ n. 摩擦
frill /frɪl/ v. 起边皱
fresh /freʃ/ adj. （指颜色）鲜明的，未褪色的
gabardine /ˈgæbədiːn/ n. 一种宽松的长袍
gown /gaʊn/ n. 女长服；礼服

leather /ˈleðə/ n. 皮革
linen /ˈlɪnɪn/ n. 亚麻布
luxurious /lʌgˈʒʊərɪəs/ adj. 十分舒适的；奢侈的
moisture /ˈmɔɪstʃə/ n. 湿气；潮湿
nylon /ˈnaɪlɒn/ n. 尼龙
polyester /ˌpɒlɪˈestə/ n. 聚酯纤维
print /prɪnt/ n. 印花布
seersucker /ˈsɪəsʌkə/ n. 泡泡纱
shed /ʃed/ v. 去掉，除掉
shrink /ʃrɪŋk/ v. 收缩；萎缩
silk /sɪlk/ n. 丝，丝绸
twill /twɪl/ n. 斜纹织物
uniform /ˈjuːnɪfɔːm/ adj. 统一的
wrinkle /ˈrɪŋkl/ n. 皱褶
wool /wʊl/ n. 羊毛；毛料

Unit 5 Suitable Clothing Fabric

Reading Comprehension

1. **Read the passage again and decide if the statements are True (T) or False (F).**

   T F
   - ☐ ☐ 1) Cotton and silk are used to make low to medium priced clothes.
   - ☐ ☐ 2) Wool is used for high-quality garments.
   - ☐ ☐ 3) People like wearing man-made fiber fabric clothes because they are cheap, colorful.
   - ☐ ☐ 4) A good designer shouldn't think much about fabrics.
   - ☐ ☐ 5) Fabrics can help inspire designers.

2. **Read the passage again and fill out the form. Try to add some new things you know.**

Fabrics	Used to make	Advantages（优点）	Disadvantages（缺点）

Speaking

 1. Here are some swatch fabrics. Talk with your partner about what kinds of clothes they can be used to make.

2. Tell your partner the fabrics you have chosen for your sample garment and invite your partner to give you some advice about them.

Language study

1. Choose the correct words or expressions in the box to fill in the blanks.

| high-priced | easy-shrinking | hard to store | warm-weather |
| high-quality | man-made | easy-frilling | |

1) This kind of _____ fiber fabric is much cheaper than the silk one.

2) Cotton is often used as _____ fabrics because it is light.

3) Cotton is a kind of _____ and _____ fabric.

4) Leather is _____ and _____.

5) Wool is often used to make _____ garments.

Unit 5 Suitable Clothing Fabric

2. Complete the sentences with the words or expressions you have learned.

1) Fabrics _____ an important role in the style of garment.

2) Fabrics can be _____ into cotton, linen, wool, silk and so on.

3) _____ man-made fabric has poor friction, it is _____ because it is colorful, soft, durable and comfortable.

4) _____ is high-priced and hard to store. It requires a lot of _____.

5) Silk is _____ weight.

3. Put the words and expressions in the correct order to make sentences.

1) first time, This is, a fabric market, in, my, to shop

   _____.

2) a kind of, It, looks like, to me, seersucker, almost

   _____.

3) recognize, of fabrics, How do you, the quality

   _____.

4) rather hard, or clean, Wool is, to wash

   _____.

5) and winter garments, applied to, Leather is, usually, woman's garments

   _____.

Acting it out

Make a garment with the fabrics including lining, trims, etc. But before that you need to discuss with your partner to complete the fabric sheet.

Fabrics you choose to make the garment.	Used in what parts of the garment.

Reading for more

FABRIC PREPARATION

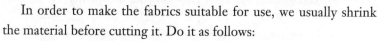

In order to make the fabrics suitable for use, we usually shrink the material before cutting it. Do it as follows:

1. Spread the fabrics' wrong side up on a flat worktable.
2. Place an L-shaped square on the fabric. Align one side of the square with a selvedge (镶边).
3. Draw a chalk line along the other side of the square so that the line is at a right angle to the selvedge.
4. Cut along the chalk line.
5. Repeat at the opposite end of the fabric.
6. If the fabric is washable, pre-shrink it by immersing (浸入) it in cold water for about two hours. Then gently squeeze (挤出) out the water, but do not wring (拧).
7. If the fabric is not washable, use a steam iron on the wrong side of the fabric, start ironing at the selvedge until you have ironed the entire length of the fabric.
8. Put the material into an airy place to dry.

Unit 5 Suitable Clothing Fabric

Self-check

In this unit I've remembered the following words:

- ☐ cotton
- ☐ polyester
- ☐ nylon
- ☐ seersucker
- ☐ denim
- ☐ twill
- ☐ gown
- ☐ brocade
- ☐ chiffon
- ☐ firm
- ☐ uniform
- ☐ print
- ☐ even
- ☐ dye
- ☐ fresh
- ☐ shed

- ☐ wrinkle
- ☐ crush
- ☐ linen
- ☐ wool
- ☐ silk
- ☐ fiber
- ☐ leather
- ☐ frill
- ☐ shrink
- ☐ conduct
- ☐ moisture
- ☐ gabardine
- ☐ elegant
- ☐ luxurious
- ☐ fade
- ☐ friction

I now understand and can use the following sentences:

_____

_____

Unit 6

Sewing Techniques

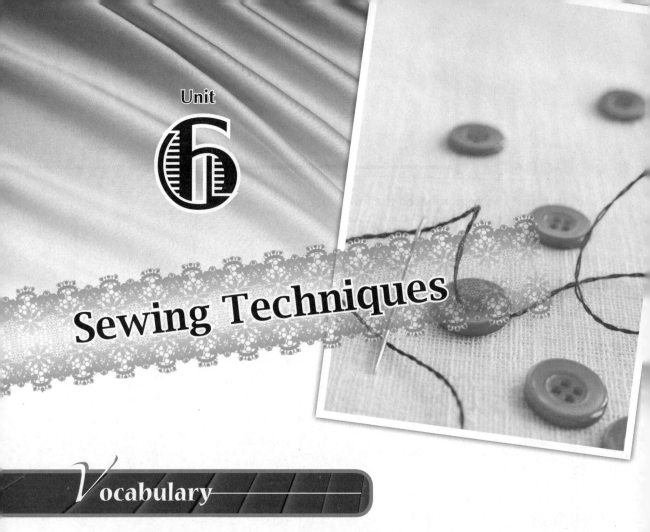

Vocabulary

1. Listen and choose the correct term in the box for each picture, change the form if necessary.

What stitches are they?

| catch stitch | blind stitch | machine stitch |
| slip stitch | back stitch | running stitch |

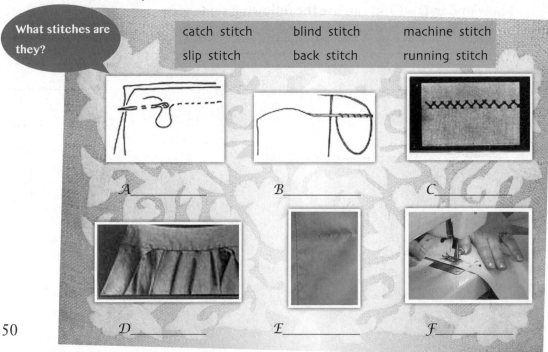

A._____   B._____   C._____

D._____   E._____   F._____

50

Unit 6 Sewing Techniques

2. Read the terms again and complete the form.

Chinese	English expression	Picture
回针针迹		
之字形针迹		
初缝针迹		
暗针针迹		
暗缝针迹		

Listening

(Li Lin visited a workshop at a college last week. She saw several students practicing hand stitching ...)

Li Lin: Excuse me, is this a special workshop? Why aren't there any sewing machines?

Student: Well, this workshop is for hand stitching only. You see the students here are all practicing hand stitching.

Li Lin: Is hand stitching more important for making clothes than sewing machine stitching?

Student: Not really. But to some degree, some sewing techniques are best done by hand. These include basting, decorative stitching, tacking and hemming.

Li Lin: Do you mean that both hand stitching and machine stitching are necessary techniques for making clothing?

Student: Yes. Step this way, please. My classmates are practicing machine stitching in this room.

Li Lin: How skillful they are!

Student: Thank you. You see, they are using machines to stitch permanent seams or finished edges, such as collar edges and front opening edges.

Li Lin: I see. I think I've learned a lot today. Thank you very much.

1. **Listen to the dialogue and tick the kinds of stitches mentioned in it.**

   ☐ machine stitching
   ☐ permanent stitching
   ☐ hand stitching
   ☐ temporary stitching

2. **Listen to the dialogue again and decide if the following sentences are True (T) or False (F).**

   T   F
   ☐   ☐   1) Hand stitching is as necessary as machine stitching.
   ☐   ☐   2) Both hand stitching and machine stitching are quite necessary for making clothing.
   ☐   ☐   3) The students learn hand stitching only.
   ☐   ☐   4) Machine stitching is often used in basting and decorative stitching.

3. **Listen to the dialogue again and complete the sentences with the expressions in the box.**

   > Ⓐ How skillful
   > Ⓑ are best done
   > Ⓒ hand stitching only
   > Ⓓ Do you mean that
   > Ⓔ a special workshop

   1) Excuse me, is this _____?
   2) This workshop is for _____.
   3) Some sewing techniques _____ by hand.
   4) _____ both hand stitching and machine stitching are necessary techniques for making clothing?
   5) _____ they are!

52

Unit 6 Sewing Techniques

Look at the following pictures. Discuss the question with your partner.
What stitches are they?

 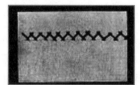 

 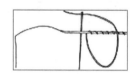

Hand Stitches

We all know that machine stitches and hand stitches are two different but necessary sewing techniques for making good clothing. Machine stitches are permanent stitches and used to hold fabrics and parts of a garment together. Hand stitches are used where machine stitches are not desirable. These include basting, decorative stitch, hemming, etc. They are important and useful in our work. Let's learn some of them.

1. **Running stitch** It is a basic hand stitch. Insert the needle from the back side of the fabric and take several stitches at a time; draw the thread through the fabric and repeat. A running stitch is always used for gathering, mending and fitting.
2. **Back stitch** It is the strongest of the hand stitches. Bring the needle through the fabric and take a small back stitch, bringing the needle out ahead of the preceding stitch. Take another back stitch, putting the needle in at the end of the preceding stitch. A back stitch is often used for under stitching where machine stitching would be difficult.
3. **Slip stitch** It is used to finish hems and tack facing almost invisibly. Bring the needle through the fold of the hem and pick up a thread of fabric at the same point.
4. **Catch stitch** It is used to finish hems on fabrics that don't fray. Work from left to right, making tiny stitches on the hem, then on the garment. Keep the stitches

loose. A catch stitch can also be used as a hemming stitch for lined garments. The blind catch stitch is hidden between garment and hem. It is quite good for knits because it remains flexible.

5. **Blind stitch** It is worked between the hem and the garment. Sew through the edge of the hem; take up one or two threads of fabric. Small stitches are almost invisible on the right side.

Words & Expressions

baste /beɪst/ v. 用长针脚疏缝；粗缝
blind /blaɪnd/ adj. 隐蔽的；视而不见的
decorative /ˈdekərətɪv/ adj. 装饰性的
edge /edʒ/ n. 边；边线；边缘
flexible /ˈfleksɪbl/ adj. 有弹性的；柔韧的；易弯曲的
fold /fəʊld/ v. 褶痕；褶缝；褶裥
fray /freɪ/ n. 磨损
gather /ˈɡæðə/ v. 给（衣服）打褶
hem /hem/ n.（裙子、窗帘等的）下摆，褶边
　　　　 v. 给…镶边
hide /haɪd/ v. 遮挡；把…藏起来
insert /ɪnˈsɜːt/ v. 插入；嵌入
invisible /ɪnˈvɪzəbl/ adj. 看不见的

knit /nɪt/ v. 编织；针织
loose /luːs/ adj. 稀松的；未织牢的
mend /mend/ v. 缝补；修补
needle /ˈniːdl/ n. 缝衣针
permanent /ˈpɜːmənənt/ adj. 持久的；永久的
preceding /prɪˈsiːdɪŋ/ adj. 前面的
stitch /stɪtʃ/ n. 线迹；针脚
　　　　 v. 缝，缝合
tack /tæk/ v.（通常指在细缝之前用大针脚）粗缝
thread /θred/ n. 线
tiny /ˈtaɪni/ adj. 极小的；微小的

at a time 每次

Reading Comprehension

1. Read the passage again and decide if the statements are True (T) or False (F).

T　F

☐　☐　1) Catch stitch is used to finish hems on fabrics that don't fray.

☐　☐　2) Blind stitch is a basic hand stitch and is always used for gathering, mending.

☐　☐　3) Running stitch is worked between the hem and garment.

☐　☐　4) Back stitch is the strongest of the hand stitches.

☐　☐　5) Slip stitch is used to finish hems and tack facing almost invisibly.

2. Read the passage again and complete the form.

Hand stitches	How to do	Characteristics

Speaking

 1. Talk about the stitches you've learned in the fashion course with your partner.

2. Tell your partner: which technique do you use when making clothing, and in what part.

Language study

1. Choose the correct expressions in the box to fill in the blanks.

| to some degree | make several stitches | why aren't there |
| keep ... loose | hold ... together | |

1) Excuse me, _____ any sewing machines in this workshop?

2) _____, some sewing techniques are best done by hand.

3) Machine stitches are permanent stitches and used to _____ fabrics and parts of garment _____.

4) Insert the needle from the wrong side of the fabric and _____ at a time.

5) Make tiny stitches on hem, then on garment. Don't forget to _____ the stitches _____.

2. Complete the sentences with the words or expressions you have learned.

1) Blind stitch is worked _____ the hem and the garment.

2) A catch stitch is quite good for knits because it remains _____.

3) Small blind stitches are almost _____ on the right side.

4) Hand stitches are used where _____ stitches are not desirable.

5) Both the hand stitch and machine stitch are necessary _____ for making clothing.

3. Choose the correct sentence in the box for each picture.

1) Back stitch is often used for under stitching where machine stitching would be difficult.

2) Running stitch is always used for gathering, mending and fitting.

3) Catch stitch is used to finish hems on fabrics that don't fray.

Unit 6 Sewing Techniques

4) Slip stitch is used to finish hems and tack facing almost invisible.

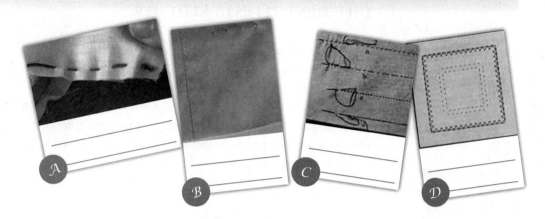

Acting it out

1. Show the garment you began to make when learning unit 5 to your partner. Introduce the stitches you used. And invite your partner to give you some advice on the troubles or problems you had in making the garment. The following sentences may be helpful.
    a. I think the back stitch here is too large.
    b. The stitches here are not connected well. You'd better do it this way.
    c. I think you'd better make a small stitch in the hem.
    d. Do you think it will be OK if you …?

2. Ask several students to give comments on your stitches.
    ☐ Well-done.   ☐ Fair.   ☐ Bad.

Reading for more

DIFFERENT TYPES OF STITCHES

There are many types of stitches used in tailoring and dressmaking, such as temporary stitches, permanent stitches, hand stitches and machine stitches.

Temporary stitches are usually used to hold the pieces and parts of a garment together temporarily, or to transfer the fitting lines from patterns onto material. They should be removed when the garment is finished.

Permanent stitches are used where the stitches are necessary. They are those which hold parts of a garment together.

Hand stitches are made by hand. They are used where machine stitches are not desirable.

Machine stitches are usually made by sewing machines, sometimes they are made by other machines. They are used to stitch permanent seams or finished edges.

Self-check

In this unit I've remembered the following words and expressions:

- ☐ stitch
- ☐ baste
- ☐ decorative
- ☐ tack
- ☐ hem
- ☐ permanent
- ☐ insert
- ☐ needle
- ☐ thread
- ☐ gather
- ☐ mend
- ☐ preceding

- ☐ invisible
- ☐ fold
- ☐ fray
- ☐ tiny
- ☐ loose
- ☐ blind
- ☐ hide
- ☐ knit
- ☐ flexible
- ☐ edge

- ☐ at a time

I now understand and can use the following sentences:

_____

_____

Unit 7

Fashion Accessories

Vocabulary

1. **Listen and choose the correct word or expression in the box for each picture, change the form if necessary.**

What are they?

earrings	sunglasses	handbag	scarf	belt
hat	bracelet	finger ring	necklace	cap
gloves	sandals	boots	watch	high-heels

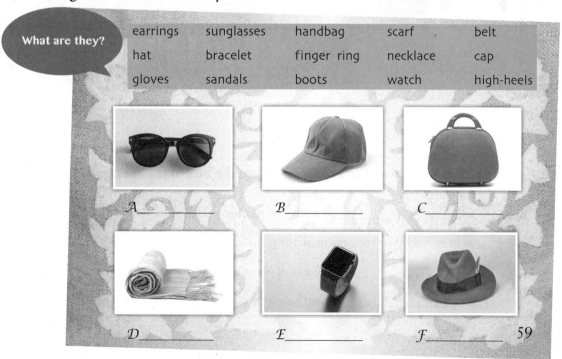

A_____ B_____ C_____

D_____ E_____ F_____

2. **Use the correct word or expression from the box to fill in each blank.**

   1) A _____ is a strip of leather or cloth that you fasten around your waist to hold your trousers or skirt up.

   2) _____ are pieces of jewelry that you wear on your ear lobes (耳垂).

   3) A _____ is used mainly by women to carry money and personal items.

Unit 7 Fashion Accessories

4) The _____ of your shoe or sock is the part that fits over the back part of your foot.

5) A _____ is a chain or band worn around someone's wrist as an ornament (装饰物).

Listening

Shop assistant:	Can I help you?
Linda:	Yes. I'm looking for a handbag.
Shop assistant:	Come this way, please. We have many kinds of handbags.
Linda:	Wow. What pretty bags! I can't decide which one I should take. Can you give me some suggestions?
Shop assistant:	I'm very glad to. I think this small hand-held bag is very suitable for a petite (娇小的) girl like you. It looks very cute.
Linda:	Really? What do you think of this handbag, Jack?
Jack:	I like the pattern, too. But why not try this hobo (流浪汉) bag? It's another style. This slouchy (松的), rounded hobo bag might compliment your figure.
Linda:	You mean I should choose a shape that is the opposite of my body type?
Jack:	That's right.
Linda:	(*trying on the bag*) Wow, this is amazing!

1. **Listen to the dialogue and tick the right picture to answer the question.**

   Which bags are suggested?

          A   B   C

2. **Listen to the dialogue again and decide if the following statements are True (T) or False (F).**

   T   F
   - ☐ ☐ 1) Linda is going shopping for a handbag.
   - ☐ ☐ 2) There are not too many handbags in the shop. So Linda can't find a satisfactory one.
   - ☐ ☐ 3) The shop assistant gives Linda some suggestions.
   - ☐ ☐ 4) Jack doesn't like the small handbag.
   - ☐ ☐ 5) Linda bought the hobo bag at last.

3. **Listen to the dialogue again and complete the sentences with the expressions in the box.**

   > Ⓐ   What pretty
   > Ⓑ   very cute
   > Ⓒ   a handbag
   > Ⓓ   Why not try
   > Ⓔ   which one I should take

   1) I'm looking for _____.
   2) _____ bags!
   3) I can't decide _____.
   4) It looks _____.
   5) _____ this hobo bag?

Unit 7 Fashion Accessories

Reading

Before You Read

Look at the following pictures. Discuss the question with your partner.
What fashion accessories do you think can be used to match these clothes?

A

B

C

Fashion Accessories

Nowadays, fashion accessories have become more and more popular with people. What accessories are hot this year?

1. Jewelry
These glittering items are the most well-known fashion accessories for all ages. Even teens and children are fond of wearing these accessories because of their bright colors. Men wear their accessories such as watches and rings. Women wear necklaces, earrings and bracelets to make them more beautiful and attractive.

2. Handbags
Handbags are also a popular fashion accessory for most women and teenage girls. Since most of them need to bring many things to their office or school, they use handbags to carry their stuff. Besides, they also need a handbag to carry their make-up to keep them fresh and beautiful all day long.

3. Shoes and Sandals
Shoes or sandals are also one of the most popular fashion accessories among women. In order to complement their different clothes, most of them own numerous pairs of

shoes in different styles and colors.

Wearing fashion accessories is indeed a good way of transforming your ordinary appearance into a surprising look. However, not all accessories work best with your clothes and your figure. So you must be careful in choosing your accessories because they can also look awkward if they do not complement your outfit and body type. Just learn how to accessorize and you'll shine best with your chosen fashion accessories.

Words & Expressions

accessory /ək'sesərɪ/ n. 配饰
appearance /ə'pɪərəns/ n. 外观；容貌
bracelet /'breɪslɪt/ n. 手镯
complement /'kɒmplɪmənt/ v. 搭配；补足
earring /'ɪərɪŋ/ n.（常用复数）耳环；耳饰
glittering /'ɡlɪtərɪŋ/ adj. 闪光的；闪耀的
indeed /ɪn'diːd/ adv. 的确；实在
jewelry /'dʒuːəlrɪ/ n. 珠宝；首饰

necklace /'neklɪs/ n. 项链
outfit /'aʊtfɪt/ n. 整套服装
sandal /'sændl/ n. 凉鞋
shine /ʃaɪn/ v. 出众；超群
stuff /stʌf/ n. 资料；东西

body type 体型

Reading Comprehension

1. Read the passage again and match the sources with the information.

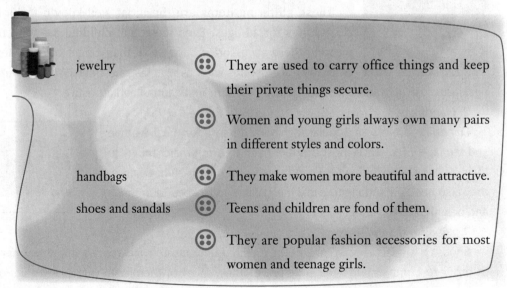

jewelry

handbags

shoes and sandals

- They are used to carry office things and keep their private things secure.
- Women and young girls always own many pairs in different styles and colors.
- They make women more beautiful and attractive.
- Teens and children are fond of them.
- They are popular fashion accessories for most women and teenage girls.

Unit 7 Fashion Accessories

2. Read the passage again and decide if the statements are True (T) or False (F).

T   F

☐   ☐   1) People use fashion accessories to complement their clothes.
☐   ☐   2) Women wear jewelry to make them more beautiful and attractive.
☐   ☐   3) Handbags are the most well-known fashion accessories for women of all ages.
☐   ☐   4) Wearing good fashion accessories is the only way to make you attractive.
☐   ☐   5) You should consider your outfit and body type when choosing your accessories.

Speaking

1. Point out the accessories your partner is wearing today and tell him / her which accessories you like best.

2. Tell your partner where you can buy the accessories.

3. Invite your partner to give you some advice on how to make the accessories work with your dress.

Language study

1. Choose the correct expressions in the box to fill in the blanks.

such as	work best with	popular with	turn … down	are fond of	because of

1) Children _____ wearing these accessories _____ the bright colors.
2) Men wear these accessories _____ watches and rings.
3) However, not all accessories _____ your clothes and your figure.
4) Some accessories can also _____ you _____ if they do not complement your outfit and body type.
5) Fashion accessories have become more and more _____ people.

2. Complete the sentences with the words or expressions you have learned.

1) What accessories are _____ this year?

2) They need a handbag to keep them _____ fresh and beautiful all day _____.

3) Shoes or sandals are one of the most popular fashion _____ among women.

4) Learn how to accessorize and you'll _____ best with your chosen fashion accessories.

5) Wearing fashion accessories is indeed a good way of _____ your ordinary appearance into a surprising look.

3. Put the words and expressions in the correct order to make sentences.

1) are also fond of, Even teens, these accessories, wearing

   _____.

2) make-up, Women also, their, to keep, need a handbag

   _____.

3) to carry, They, handbags, have to, their stuff, use

   _____.

4) in choosing, You, your accessories, must, be careful

   _____.

Acting it out

1. Collect all of the accessories from you and your partner together.

2. Have a talk with your partner about the accessories and choose some for your sample garment finished during units 5-6.

3. Put on the garment with different accessories and show it to your class. Try to get some advice from your classmates.

Reading for more

WENDY'S FASHION ACCESSORIES WEBSITE

| WATCHES | MEN'S | LADIES' | JEWELRY | HANDBAGS | SHOES | SUNGLASSES | NEW ARRIVALS |

Style	Occasion	Size	Material	Best Sellers
Backpacks	Bridal	Mini	Canvas	Key-chains
Clutches	Casual	Small	Cotton	Drawstring
Cross-body	Cocktail	Medium	Leather	Leather Satchels
Messengers	Formal	Large	Nylon	
Satchels	Sport	Extra-Large	Suede	
Shoulder	Travel			
Totes	Work	**Features**	**Accessories**	
Wallets		Adjustable	Bag Charms	
Briefcases		Laptop Pocket	Cosmetic Cases	
		Silver-tone	Passport Holders	
			Travel Cases	
View All	**View All**			

SALE

fashion

Sign in

New customers? Start here.

Self-check

In this unit I've remembered the following words and expressions:

- ☐ accessory
- ☐ earring
- ☐ bracelet
- ☐ necklace
- ☐ sandal
- ☐ jewelry
- ☐ glittering
- ☐ stuff

- ☐ complement
- ☐ indeed
- ☐ appearance
- ☐ outfit
- ☐ shine

- ☐ body type

I now understand and can use the following sentences:

_____

_____

Unit 8

Fashion Show Planning

Vocabulary

1. Listen and choose the correct word or expression in the box for each picture, change the form if necessary.

What can they inspire you?

rehearse   model   lighting   fashion show

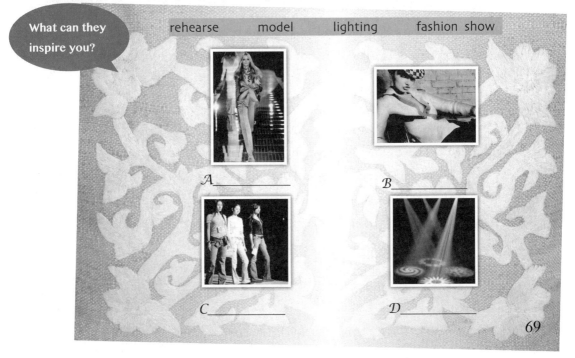

A._____   B._____

C._____   D._____

69

2. **Use the correct word or expression from the box to fill in each blank.**

   1) _____ means light of a particular type or quality.

   2) A _____ is an event at which some persons show new styles of clothes.

   3) A _____ is someone whose job is to show clothes, make-up etc.

   4) The word "_____" means practicing a play, concert, opera, etc. before giving a performance.

Listening

Li Hong:	The fashion show is coming up. It's for the graduation ceremony. I think it is very important for us all.
Peter:	I agree. So it's time to make a specific plan now.
Li Hong:	Let's write down what preparations we have to do at once.
Peter:	Sure. The theme of the show should be decided first.
Li Hong:	Yes, and then the collections, accessories, stage, lighting and the sound.
Peter:	But don't forget the models. Who can be the models?
Li Hong:	It's a piece of cake. Why not invite some students from the drama department to be the models.
Peter:	Good idea. But …
Li Hong:	What's the matter?
Peter:	You know, the walking, timing, posing and turning are all very important aspects of a model's presentation on the runway. I'm afraid they are short of such experience.
Li Hong:	Don't worry. I know a very experienced teacher. He can help them rehearse their routines.
Peter:	Great. Let's get going.

Unit 8 Fashion Show Planning

1. **Listen to the dialogue and tick the right picture to answer the question.**
   What is Peter afraid of during making the plan of the show?

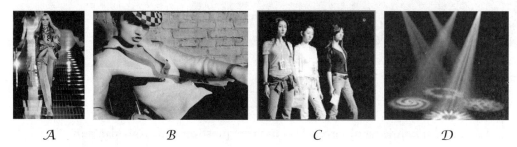

   A   B   C   D

2. **Listen to the dialogue again and decide if the following statements are True (T) or False (F).**

   T   F
   □   □   1) There is going to be a fashion show before Peter's graduation.
   □   □   2) Peter and Li Hong have got everything ready.
   □   □   3) Peter and Li Hong will be models.
   □   □   4) An experienced teacher will help the students from the drama department to rehearse.
   □   □   5) Walking, timing, posing and turning are very important for a model's presentation.

3. **Listen to the dialogue again and complete the sentences with the expressions in the box.**

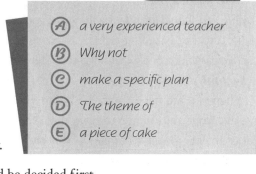

   Ⓐ a very experienced teacher
   Ⓑ Why not
   Ⓒ make a specific plan
   Ⓓ The theme of
   Ⓔ a piece of cake

   1) It's time to _____ now.
   2) _____ the show should be decided first.
   3) It's _____.
   4) _____ invite some students from the drama department to be the models?
   5) I know _____. He can help them rehearse their routines.

71

Reading

Before You Read

Look at the following pictures. Discuss the question with your partner.
What preparations must be done for a fashion show?

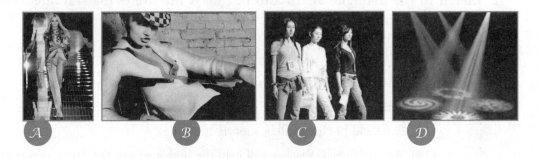

A   B   C   D

Planning a Fashion Show

Have you ever got the idea to put on your own fashion show? Working through this topic, you'll find it is not so difficult!

First, make a plan.

The plan should include the theme of your show, collection, models, accessories & props, venue, catwalk & stage set, lighting, and sound. If you can consider all these aspects, you will have a clearer idea of how to stage your show.

Second, discuss your plans.

When you are happy with your plan, discuss it with others involved in the planning, give everyone the chance to have their say. Then decide who will do what, make sure that all the people are clear about what their roles are.

Third, get on with the job.

Models not only have good posture, but also convey the theme of your show. Get help from specialists, such as a performing arts teacher — who will be able to help the models rehearse their routines, a lighting technician, a sound technician, an emcee, a

Unit 8 Fashion Show Planning

stagehand, a speaker, etc.

Fourth, bring it all together.

You will need at least three full rehearsals, including one final dress rehearsal to bring the show together and make it look professional.

Finally, check everything that needs to be in place before the show begins.

In this topic you have looked at what's involved in planning a fashion show. Now it's time to put it into practice and produce a show-stopper of a fashion show!

Words & Expressions

collection /kə'lekʃn/ n. 时装展销
convey /kən'veɪ/ v. 传达；表达
emcee /ˌem'siː/ n. 司仪；节目的主持人
posture /'pɒstʃə/ n. 姿态
prop /prɒp/ n. 道具
professional /prə'feʃənl/ adj. 专业的；内行的
routine /ruː'tiːn/ n. 表演的成套动作
rehearse /rɪ'hɜːs/ v. 排练
show-stopper n. （被长时间的掌声所打断的）精彩表演

specialist /'speʃəlɪst/ n. 专业工作者
stagehand /'steɪdʒhænd/ n. 舞台工作人员
theme /θiːm/ n. 主题
venue /'venjuː/ n. 会场；地点

dress rehearsal 带妆彩排
performing arts 表演艺术
stage set 舞台背景

Reading Comprehension

1. Read the passage again and decide if the statements are True (T) or False (F).

T F

☐ ☐ 1) You should make a plan before staging your show.
☐ ☐ 2) It's not necessary for people to know what you want them to do.
☐ ☐ 3) It's OK to choose models who only have good posture.
☐ ☐ 4) A performing arts teacher can help the models rehearse their routines.
☐ ☐ 5) If you want things to run smoothly, make sure everything is in place before the show begins.

2. Read the passage again and put the phrases in correct order according to the passage.

   How do you plan a fashion show?

   _____ select the models

   _____ arrange rehearsals

   _____ make sure everything is in place

   _____ find some specialists for help

   _____ think about how to present your show

   _____ discuss the idea plan with others

Speaking

1. Have you seen a fashion show on TV or somewhere? Please tell your partner what is needed for a fashion show.

2. Suppose you are going to plan a fashion show at school. Talk with your partner about the preparations you need to do and troubles you might have and find the solutions.

74

Language study

1. Choose the correct expressions in the box to fill in the blanks.

| involved in | put on | in place | put … into practice | have … say |

1) Have you got an idea to _____ your own fashion show?

2) Now it's time to _____ it _____ and produce a show-stopper of a fashion show.

3) When you are happy with your plan, discuss it with others _____ the planning.

4) Please give everyone the chance to _____ their _____. Then your plan can be made better.

5) Check everything that needs to be _____ before the show begins.

2. Complete the sentences with the words or expressions you have learned.

1) _____ through this topic, you'll find it is not so difficult planning a fashion show.

2) Make sure all the people _____ what their roles are.

3) You will need three full rehearsals, _____ one final dress rehearsal to bring the show together.

4) Models not only have good posture, but also _____ the theme of the show.

5) The plan should _____ the theme of your show, collection, models, accessories & props, venue, catwalk & stage set, lighting, and sound.

3. **Choose the correct sentence for each picture.**

   1) Getting help from a sound technician is quite necessary for a fashion show.
   2) Good lighting is necessary to show clothing to its best advantage.
   3) At least three full rehearsals are needed including one final dress rehearsal.
   4) Models not only have good posture, but also convey the theme of the show.

Acting it out

Suppose the *Fashion Week* of your school is coming. You and your classmates are invited to take part in the fashion show. Your garments made during the period of studying *Fashion English* will be shown. Please talk with your group members about the preparations you need to do for the show, especially the arrangement of the stage, lighting, music, the choosing of the models, etc. Write down the troubles you have and try to find the solutions.

Preparations	Troubles or problems	Solutions

Reading for more

TIPS ON GOING FOR A CASTING

As a professional model, you may have various chances to go for a casting. Here are some helpful tips for you:

1. Make a model card beforehand. The model card should include the information as follows: your name, age, nationality, height, weight and your vital statistics.

2. You have to be confident. Although people's face, body and other conditions are natural, your stage display can also affect the interviewer. Don't think about your body defects, make-up imperfections and many other negative factors. Try to stand out in a crowd.

3. Be punctual for your casting. No brand or designer is willing to use the models who are always late, no matter how popular he or she is.

4. Choose the clothes that can show your best figure. Tights and black suspenders can show the whole shoulder, highlighting youth, vitality and good figure; a short dress without showing too much skin on your upper body can show your enviable long legs. Never wear the same clothes as your competitors.

5. Don't wear heavy make-up. Heavy make-up will determine the model's style, while light make-up is full of possibilities, stimulating designer's imagination.

6. Show your professional catwalk. When a group of models walk together, whose steps are stable and whose training is professional can be noticed clearly on the T-stage.

7. Try to maintain good eye contact with the designers. When the walk show is over, there will be a static casting. Observe the designers under the stage carefully. They will take photos of the models they may hire. Strike a pose when you are photographed.

8. Cooperate with agents actively. The model agents will teach you a lot from a professional point of view and provide you opportunities.

Above all, you need to receive professional training because the lack of professional skills would have side effects. So if you want to become an excellent model, the first thing you need to do is to receive professional training.

Self-check

In this unit I've remembered the following words and expressions:

- [ ] theme
- [ ] collection
- [ ] prop
- [ ] venue
- [ ] specialist
- [ ] routine
- [ ] stagehand
- [ ] rehearse
- [ ] emcee

- [ ] posture
- [ ] convey
- [ ] professional
- [ ] show-stopper

- [ ] stage set
- [ ] dress rehearsal
- [ ] performing arts

I now understand and can use the following sentences:

_____

_____

Unit 9

Garment Marketing

Vocabulary

1. Listen and choose the correct word or expression in the box for each picture, change the form if necessary.

What are needed in garment marketing?

retailer	high-quality	salesperson
discount	merchandising policies	consumer psychology
workmanship	efficient communication	image

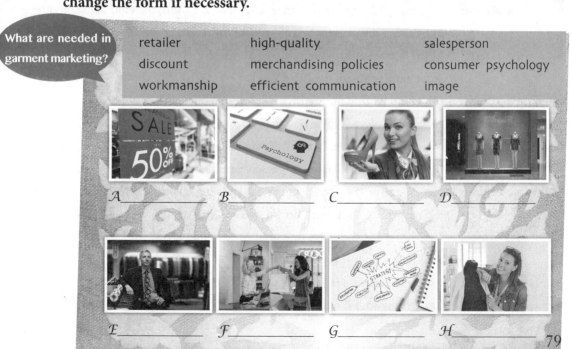

A_____ B_____ C_____ D_____

E_____ F_____ G_____ H_____

2. **Use the correct word or expression from the box to fill in each blank.**

   1) A _____ is a reduction in the price of something.

   2) A _____ is someone who buys and uses goods and services.

   3) A _____ is a set of ideas or plans that is used as a basis for making decisions, especially in politics, economics or business.

   4) A _____ is a person or business that sells goods to the public.

   5) _____ is the opinion of yourself, your company, or your community that you try to create in the minds of other people.

Listening

Shop assistant:	Good afternoon! Can I help you?
Mrs Green:	Yeah. I need proper clothing for an important party. Could you give me some suggestions, please?
Shop assistant:	Of course, it's my honor. Since it is a party, I think the clothing should be fashionable and elegant. Would you like the model gown? It is our best seller.
Mrs Green:	Yeah. It's beautiful, but I don't like the color or the style. Do you have something more unique?
Shop assistant:	Oh, I see. How about this one? It's fashionable, simple and elegant. The design is very special.
Mrs Green:	Wow! I like the style, but I don't care for the color, and it is a bit small, too. Have you got any larger ones?
Shop assistant:	Yes. We have this model in several sizes and colors, such as white, black, dark brown, pink …
Mrs Green:	Let me try on the white one in my size. (*After trying it on.*) How does it look?
Shop assistant:	You look beautiful in it. It fits you well.
Mrs Green:	Yes, I think so. I'll take this one.

Unit 9   Garment Marketing

1. Listen to the dialogue and tick the right picture to answer the question.
   What clothing will Mrs Green get?

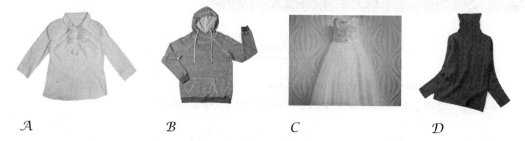

   A   B   C   D

2. Listen to the dialogue again and decide if the following sentences are True (T) or False (F).

   T   F
   ☐   ☐   1) Mrs Green wants to buy the dress for an important meeting.
   ☐   ☐   2) Mrs Green doesn't like the color of the model gown but prefers its style.
   ☐   ☐   3) The dress is too long and loose for Mrs Green.
   ☐   ☐   4) The shop assistant thinks the design of the dress is special.
   ☐   ☐   5) Mrs Green took the white one at last.

3. Listen to the dialogue again and complete the sentences with the expressions in the box.

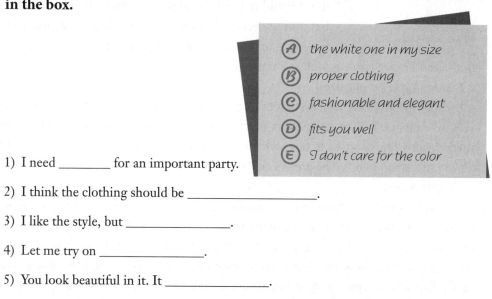

   Ⓐ the white one in my size
   Ⓑ proper clothing
   Ⓒ fashionable and elegant
   Ⓓ fits you well
   Ⓔ I don't care for the color

   1) I need _____ for an important party.

   2) I think the clothing should be _____.

   3) I like the style, but _____.

   4) Let me try on _____.

   5) You look beautiful in it. It _____.

81

Reading

Before You Read

Discuss the question with your partner.
If you're shopping for clothes, what are the main factors in your decision?

☐ the styles ☐ the price ☐ the quality ☐ the brand ☐ other factors

Merchandising Tactics

Merchandising tactics employed by retailers are very useful to maintain and attract customers. Among the many merchandising tactics, the following elements are the most important.

High-quality salespersons

Efficient communication with customers is the key to garment merchandising. As for high-quality salespersons, they should not only have professional garment knowledge, but also understand consumer psychology. They must know how to gain the trust of customers through proper wear, cordial and sincere tone, as well as careful service.

Fashion cycle emphasized

In order to form its image, every retailer decides to emphasize one phase of the fashion cycle over others. It then chooses its merchandise to fit that phase. A retailer can be a fashion leader if he chooses to buy styles in the introductory phase, or a retailer may be a follower of fashion trends if he prefers clothes in the culmination or peak stages.

Quality policy

Generally, there are three levels of quality for retailers to choose from: the top level involving the finest materials and workmanship; or the intermediate level, which shows concern for quality and workmanship but always maintains certain price levels; or the serviceable level having materials and workmanship of a very low level and keeping equally low prices. Once a store has set its quality tactics, it must make more specific decisions, such as whether it will accept nothing less than perfect goods or whether it will permit irregulars and second-quality goods to be offered.

Retailers also focus on other aspects, such as price ranges, depth and breadth of assortments, exclusivity, brand policies and so on. Regardless of the methods chosen, satisfied customers are the ultimate goal.

Words & Expressions

consumer /kən'sjuːmə/ n. 消费者
cordial /'kɔːdɪəl/ adj. 亲切的
culmination /ˌkʌlmɪ'neɪʃən/ n. 顶点；巅峰
emphasize /'emfəsaɪz/ v. 突出；强调
exclusivity /ekˌskluː'sɪvɪtɪ/ n. 独特性
intermediate /ˌɪntə'miːdɪət/ adj. 中间的
irregulars /ɪ'regjuləz/ n. 次品
maintain /meɪn'teɪn/ v. 保留
merchandise /'mɜːtʃəndaɪz/ n. 商品；货品
　　　　　　　　　　　　　v. 销售
phase /feɪz/ n. 阶段；时期
psychology /saɪ'kɒlədʒɪ/ n. 心理学
retailer /'riːteɪlə/ n. 零售商

serviceable /'sɜːvɪsəbl/ adj. 能用的；实用的
sincere /sɪn'sɪə/ adj. 诚恳的
target /'tɑːgɪt/ n. 目标
tone /təʊn/ n. 语气；音调
workmanship /'wɜːkmənʃɪp/ n. 手艺；工艺

depth and breadth of assortments 商品分类的
　　深度和广度
fashion cycle 流行周期
fashion trends 流行趋势
merchandising tactics 销售策略
model gown 模特样衣
price ranges 价格范围

Reading Comprehension

1. **Read the passage again and match the events with the merchandising policies.**

   Events

   1) The retailer selling the clothing with the finest materials and workmanship doesn't permit irregulars to appear in his store.

   2) Having finished the training, the salespersons know how to offer good service to their customers.

   3) With purple being popular, many garments with purple are sold in a clothes store.

   Merchandising policies

   - High-quality salespersons and service training
   - Fashion cycle emphasized
   - Quality policy

2. **Read the passage again and fill out the form according to the questions.**

What is the key to garment merchandising?	The key to garment merchandising is _____.
What do high-quality salespersons have?	High-quality salespersons not only _____, but also _____.
How can the salespersons gain the trust of customers?	They gain the trust of customers through _____.
Why does every retailer decide to emphasize fashion cycle?	It is because _____.
What can a retailer be if he buys new styles or popular styles?	He can be _____ if he buys new styles, while he can be _____ if he prefers to popular styles.
How many levels of quality can retailers choose from? What are they?	There are _____. They are _____.
What other policies do retailers focus on?	They also focus on _____.

Unit 9 Garment Marketing

Speaking

Suppose you are going to start an online clothing shop. Discuss with your partner to decide the merchandising polices for your shop.

Language study

1. Choose the correct expressions in the box to fill in the blanks.

| in several sizes | gain the trust of | focus on | communication with | emphasize on |

1) Efficient _____ customers is the key to garment merchandising.

2) The salesperson must know how to _____ customers through proper wear, cordial and sincere tone as well as careful service.

3) In order to form its image, every retailer decides to _____ one phase of the fashion cycle over others.

4) Retailers also _____ other aspects, such as price ranges and so on.

5) We have this model _____ and colors, such as white, black, dark brown, pink ...

2. Complete the sentences with the words or expressions you have learned.

1) Merchandising tactics employed by _____ are very useful to _____ _____ and attract customers.

2) As for _____ salespersons, they should not only have the _____

85

garment knowledge, but also _____ customer psychology.

3) There are three _____ of quality for retailers to choose from.

4) Once a store has _____ its _____ tactics, it must make more specific _____.

5) Regardless of the methods chosen, satisfied customer are the _____ goal.

3. Choose the correct sentence in the box for each picture.

> 1) If a retailer chooses to buy clothes in the culmination or peak stages, he may be a follower of fashion trends.
> 2) The salesperson who gives you a lot of help understands consumer psychology quite well.
> 3) The top quality level involves the finest materials and workmanship, so the price of this kind of clothing is very expensive.
> 4) Merchandising policies established by retailers are used widely.

A _____

B _____

C _____

D _____

86

Unit 9 Garment Marketing

Acting it out

1. Suppose you are planning to start a clothing shop online. Tell your partner about this and invite him / her to give you some advice on garment merchandising.

2. Design the homepage of your online shop with your partner.

Reading for more

TYPES OF STORES

There are many different kinds of retail operations: department stores, specialty stores, chain operations, discount stores, and leased departments, etc. Almost all retail stores used to offer some form of mail-order or telephone or fax buying service. Many retailers are setting up Internet services. Some stores have grown into large operations, but many others are still small independently owned and operated businesses.

Specialty retailers offer limited lines of related merchandise. Examples of specialty stores include shoe stores, jewelry stores and boutiques. Another variation of the specialty store is the private label retailer, which sells only what it manufacturers itself.

Discount stores sell name-brand merchandise at less than retail prices. Discounters make a profit by keeping their overhead low and service minimal. Most discount stores have centralized checkout counters and rely on self-service.

The department store is the type of general retailer most familiar to the buying public. These retailers sell many kinds of merchandise in addition to clothing. Department stores are organized into special areas or departments, such as sportswear, dresses, men's clothing, children's wear.

Self-check

In this unit I've remembered the following words and expressions:

- ☐ merchandise
- ☐ establish
- ☐ consumer
- ☐ psychology
- ☐ cordial
- ☐ sincere
- ☐ tone
- ☐ emphasize
- ☐ phase
- ☐ retailer
- ☐ maintain
- ☐ culmination
- ☐ workmanship

- ☐ intermediate
- ☐ serviceable
- ☐ irregulars
- ☐ exclusivity
- ☐ target

- ☐ model gown
- ☐ fashion trends
- ☐ merchandising tactics
- ☐ fashion cycle
- ☐ price ranges
- ☐ depth and breadth of assortments

I now understand and can use the following sentences:

_____

_____

Unit 10

Portfolio

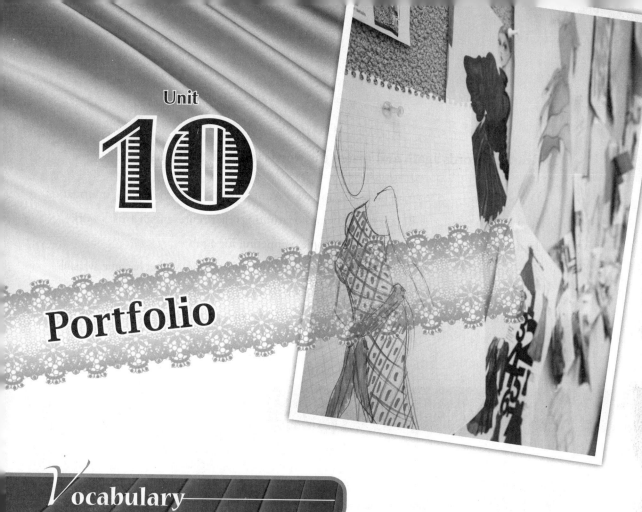

Vocabulary

1. Listen and choose the correct word in the box for each picture, change the form if necessary.

   portfolio   candidate   collection
   works   document   interviewer

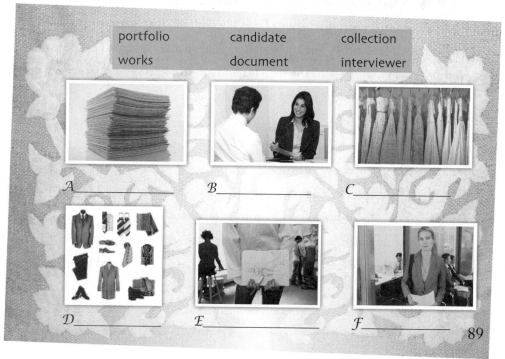

A_____ B_____ C_____

D_____ E_____ F_____

2. Read the words again and match the words to the definitions.

1) collection   A. a piece of paper or a set of papers containing official information

2) portfolio   B. a set of drawings or paintings that represent an artist's work

3) interviewer   C. one of the people competing for a job

4) candidate   D. new clothes or other products for the home that become available at a particular season

5) document   E. someone who interviews people, especially for a job

Listening

(Wang Gang, who looks unhappy, met April on his way to the library.)

April: Hi, Wang Gang. You look unhappy. What's up?

Wang: I received a phone call asking me to take part in a job interview the day after tomorrow.

April: What's wrong with you? You should be excited.

Wang: Yes. But they asked me to bring my portfolio with me that day. I don't know how to prepare for that. You know we have done so many designs. Do I need to take them all?

April: Don't worry. I heard that David had just finished an interview in a big design company yesterday. Let's ask him for some advice.

(A few minutes later.)

April: Hi! David, are you free now?

David: Sure. What's up?

Wang: I'll be interviewed in two days. Could you give me some advice?

David: Well, first choose several up-to-date works. Feel free to talk to the interviewer while showing your works to him / her. It's best for them to learn as much about you as possible in a short time.

Unit 10　Portfolio

Wang:	I see. Thank you. But I wonder how many pieces are acceptable for an interview.
David:	I think no more than twelve are suitable.
Wang:	Do I need to leave my portfolio with the interviewer after the interview?
David:	No, I don't think so. You'd better bring it back.
Wang:	Thanks very much for your help.
David:	It's my pleasure. Good luck to you.

1. Listen to the dialogue and match the sentences with the pictures.

1) Let's ask him for some advice .
2) David had just finished an interview.
3) You look unhappy. What's up?
4) You know we have done so many designs. Do I need to bring them all?

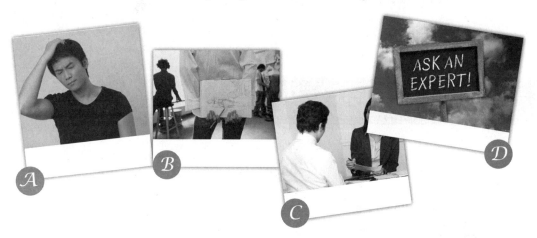

2. Listen to the dialogue again and number the sentences.

☐ But I don't know how to prepare my portfolio.
☐ You look unhappy. What's up?
☐ Let's ask David for some advice.
☐ I was asked to take part in a job interview.
☐ You should be excited.

91

3. **Listen to the dialogue again and complete the sentences with the expressions in the box.**

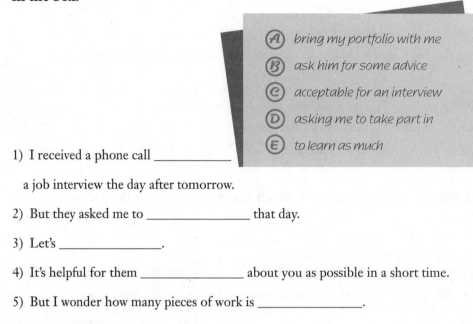

Ⓐ bring my portfolio with me
Ⓑ ask him for some advice
Ⓒ acceptable for an interview
Ⓓ asking me to take part in
Ⓔ to learn as much

1) I received a phone call _____ a job interview the day after tomorrow.

2) But they asked me to _____ that day.

3) Let's _____.

4) It's helpful for them _____ about you as possible in a short time.

5) But I wonder how many pieces of work is _____.

Before You Read

Look at the following pictures. Discuss the question with your partner.
Which portfolio is your favorite? Why?

Portfolio

Getting ready for a job interview is not an easy thing for any student, especially for one majoring in art. They not only need to prepare a well-written résumé, a suitable portfolio is also quite necessary for an interview.

Most art and design-related jobs require the candidate to present a collection of his best works. The interviewers or the employers can quickly have a look at your works and get a good idea of your drawing skills, sense of design, feeling for color, texture and your ability of coordination. So it is quite important for you to prepare a good portfolio before a job interview. Here are some tips:

1. It is not necessary to present all of your collected works. Showing eight to twelve pieces of your works to the interviewer is suitable.

2. Make sure to present your truly up-to-date works, for the employers or interviewers are too busy to spend much time going over large numbers of works. In fact, they show more interest in discovering if your ideas are a season or two ahead of the time.

3. During the interview, you should be free to ask the interviewers if they would like you to talk about your works or ideas. It may be helpful for them to know more about your abilities.

4. Keep your portfolio as neat as possible.

5. Prepare different portfolios for different job interviews, what is suitable for one position may not be acceptable for another interview.

6. Don't leave your portfolio with an employer. If you are asked to do so, arrange a return time.

7. Never borrow designs from another artist and present them as your own work! Show only your own work.

Keep these tips in mind; they are very useful and will help you get a job.

Words

acceptable /əkˈseptəbl/ *adj.* 可接受的；可允许的
candidate /ˈkændɪdɪt/ *n.* 求职者
coordination /kəʊˌɔːdɪˈneɪʃən/ *n.* 协调；调节
employer /ɪmˈplɔɪə/ *n.* 雇主；雇佣者
interview /ˈɪntəvjuː/ *n.* 面试；面谈
　　　　 v. 进行面试
interviewer /ˈɪntəvjuːə/ *n.* 面试官；接见者

major /ˈmeɪdʒə/ *n.* 主修科目
　　　 v. 主修(某一科目)
neat /niːt/ *adj.* 整齐的；整洁的
portfolio /ˌpɔːtˈfəʊlɪəʊ/ *n.* 作品集；代表作
position /pəˈzɪʃən/ *n.* 工作职位
present /prɪˈzent/ *v.* 提供；提交
résumé /ˈrezjʊmeɪ/ *n.* 简历
texture /ˈtekstʃə/ *n.* 手感；质感；质地
up-to-date /ˌʌp təˈdeɪt/ *adj.* 最新的；新近的；新式的

Reading Comprehension

1. Read the passage and decide if the statements are True (T) or False (F).

T   F

☐ ☐ 1) Collect as many works as you can to prepare a good portfolio.

☐ ☐ 2) Keep your portfolio clean and up-to-date.

☐ ☐ 3) Try to talk with the interviewers about your works.

☐ ☐ 4) Prepare one portfolio for different job interviews.

2. Read the passage again and answer the questions.

1) What do you need to prepare for a job interview?
2) How many pieces of work are needed for a good portfolio?
3) What should your portfolio be like?
4) What do the interviewers expect to see in your works?
5) If you are asked to leave your portfolio with the employer, what do you need to do then?

Speaking

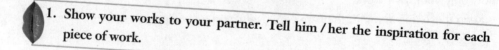

1. Show your works to your partner. Tell him / her the inspiration for each piece of work.

2. Tell your partner you want to choose some of them to prepare a portfolio, and ask for some advice.

Unit 10   Portfolio

Language study

1. Choose the correct expressions in the box to fill in the blanks.

| keep … in mind | ahead of | neat | majoring in | going over |

1) Getting ready for a job interview is not an easy thing for any student, especially for one _____ art.

2) Employers are too busy to spend too much time _____ a large number of works.

3) _____ these tips _____. They're probably good for your interview.

4) In fact, they show more interest in discovering if your ideas are a season or two _____ the time.

5) Keep your portfolio as _____ as possible.

2. Complete the sentences with the words or expressions you have learned.

1) Most art and design-related jobs require the _____ to present their own works.

2) The interviewers or the employers can quickly have a look at your works and _____ your drawing skills, sense of design, feeling for color …

3) It is not necessary to _____ all of your _____ works.

4) Bringing a résumé and a well-prepared _____ is necessary for a job interview.

5) What is _____ for one position may not be _____ for another interview.

95

3. Choose the correct sentence in the box for each picture.

1) Don't leave your portfolio with an interviewer.
2) Be free to talk with the interviewer about your works or ideas.
3) Don't present all of your collected works.
4) Keeping your portfolio neat is necessary.

Acting it out

1. Choose some up-to-date works to form your portfolio.

2. Show your portfolio in your class and invite your classmates to give you some suggestions.

3. Invite your teacher to pretend to be an interviewer. You are interviewed by your teacher. Feel free to talk to your teacher.

Reading for more

THE RÉSUMÉ

A well-designed résumé is an "interview getter". Here are a few résumé-writing tips:
1. Make use of sample résumés and guide books.
2. Spend as much time as you need organizing all of your information clearly.
3. Revise your draft until you are satisfied that you have covered all areas that may be important to share with the reader of your résumé.
4. Never misrepresent your experience or work history. Be truthful and emphasize your strong points.
5. Describe what abilities you have and what you would like to be doing in the future.
6. No need to include race or religion.
7. Don't type the word résumé at the top of the page.
8. As a beginner, your résumé should not be longer than one typewritten page. Handwritten résumés are thought of as unprofessional. Messy résumés or those with typing, spelling, or factual errors are unacceptable.

Remember your résumé is your advertisement for yourself!

Be sure that your résumé covers the following areas: Name and Address, Goal, Education, Work Experience, Unpaid Experience and Special Skills and Interests.

Self-check

In this unit I've remembered the following words:

- ☐ interview
- ☐ portfolio
- ☐ interviewer
- ☐ major
- ☐ résumé
- ☐ candidate
- ☐ present

- ☐ employer
- ☐ texture
- ☐ coordination
- ☐ up-to-date
- ☐ neat
- ☐ position
- ☐ acceptable

I now understand and can use the following sentences:

_____

_____

Unit 11

Career Expectations

Vocabulary

1. Listen and choose the correct word or expression in the box for each picture, change the form if necessary.

What jobs are these in the field of fashion industry?

designer	cutter	fashion editor
cosmetics consultant	color consultant	fabric designer
assistant designer	pattern maker	retailer
wholesaler		

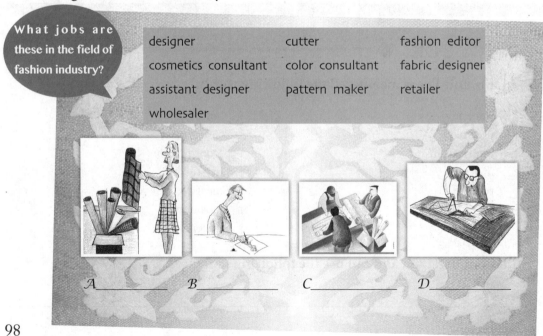

A_____ B_____ C_____ D_____

Unit 11 Career Expectations

E_____ F_____ G_____ H_____

I_____ J_____

2. Use the correct word or expression from the box to fill in each blank.

1) A _____ is a person who gives expert advice to people who need professional help.

2) _____ are substances such as lipstick or powder, which people put on their faces to make themselves look more attractive.

3) A _____ cuts the fabrics and other materials into shapes as dictated by the patterns.

4) An _____ is a person who is in charge of a newspaper or magazine.

5) A _____ is someone whose job is to sell large quantities of goods to shops or small businesses.

Listening

(Zhang Wei and his classmates are going to graduate next month. They are talking about their expectations.)
Zhang Wei: There are so many jobs to choose in the field of fashion. What are you going to do after graduation?

Yang Hong:	I think writing for a fashion magazine must be fun. I'd like to have that job.
Gao Li:	I think it is boring to write every day. I prefer to be a good pattern maker.
Zhang Wei:	Are you kidding? You have learned fashion design for four years. Why not try to be a designer?
Gao Li:	I don't agree with you. Most companies now use computers for pattern makers, but in fact few people choose to specialize in this area because they're short of a working knowledge of computer programs. However, that's what I am interested in.
Zhang Wei:	That sounds reasonable. Then do I need to change my plan for the future?

1. **Listen to the dialogue and tick the right picture to answer the question.**
   What job is mentioned in the dialogue?

   A   B   C   D

2. **Listen to the dialogue again and decide who gives the following opinions.**

Names	Whose opinions are they?
Zhang Wei Yang Hong Gao Li	A. I think writing for a fashion magazine must be fun.
	B. I think it is boring to write every day.
	C. I prefer to be a good pattern maker.
	D. That sounds reasonable.
	E. Why not try to be a designer?

Unit 11　Career Expectations

3. **Listen to the dialogue again and complete the sentences with the expressions in the box.**

- (A) what I'm interested in
- (B) change my plan
- (C) writing for a fashion magazine
- (D) so many jobs to choose
- (E) use computers for pattern makers

1) There are _____ in the field of fashion.

2) I think _____ must be fun.

3) Most companies _____ now.

4) However, that's _____.

5) Then do I need to _____ for the future?

Reading

Before You Read

Look at the following pictures. Discuss the question with your partner.
What position does each picture indicate?

A

B

C

D

101

Chances in Fashion Industry

The world of fashion is full of challenging, exciting and high salary career opportunities. In this field, it is easy to change jobs and be free to move to a different city without having to begin a new job in a different area. Fashion editors, fabric designers, color consultants, designers, wholesalers, retailers, television and Internet salespersons, marketing consultants and public relations consultants are just some of the areas of employment. Working in these fields is fun. Each area has its own requirements and an understanding of fashion is especially important in some of those jobs.

Designer: The most successful designers should have an educational background that includes sketching, draping or drafting and sewing. Besides, insight and intuition always play a big part in a designer's success. Constant experimentation with new ideas is a must. To some degree, designers are the mainstays of fashion companies. Without their creations, there wouldn't be any lines to sell.

Assistant designer: Assistant designers are responsible for the sewing of the sample garments, the selection of trimmings and the investigation of the textiles for fabrics. Besides, they must aid the designers in any way necessary.

Pattern maker: Pattern makers create the pattern according to the original design. The pattern will be finally used to produce the finished garments. Therefore they must be technically trained in construction, grading of sizes, production, cutting and fabric utilization. Most companies now use computers for pattern makers. So the pattern makers need a working knowledge of computer programs. Salaries for this position are very high because not too many people choose to specialize in this area.

Cutter: Cutters always cut the fabrics and other materials into shapes as dictated by the patterns. This career requires technical skill, including familiarity with computerized cutting.

Words & Expressions

challenge /'tʃælɪndʒ/ *n.* 挑战；比赛
consultant /kən'sʌltənt/ *n.* 顾问
editor /'edɪtə/ *n.* 编辑
employment /ɪm'plɔɪmənt/ *n.* 雇佣

familiarity /fə,mɪlɪ'ærɪtɪ/ *n.* 熟悉；通晓
insight /'ɪnsaɪt/ *n.* 洞察力
intuition /,ɪntjuː'ɪʃən/ *n.* 直觉
investigation /ɪn,vestɪ'geɪʃən/ *n.* 调查；研究

Unit 11 Career Expectations

mainstay /'meɪnsteɪ/ n. 支柱
original /ə'rɪdʒənəl/ adj. 原始的；最初的；最早的
production /prə'dʌkʃən/ n. 制造；生产
responsible /rɪ'spɒnsəbl/ adj. 需负责的

selection /sɪ'lekʃən/ n. 挑选；选择
utilization /ˌjuːtɪlaɪ'zeɪʃən/ n. 利用；应用
wholesaler /'həʊlseɪlə/ n. 批发商
be responsible for 对…负责任

Reading Comprehension

1. Read the passage again and list some of the jobs mentioned in the text.

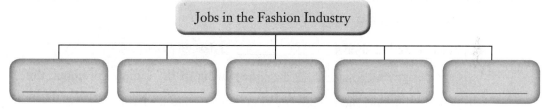

2. Read the passage again and decide if the statements are True (T) or False (F).

   T   F
   □   □   1) There are many career opportunities in the fashion industry.
   □   □   2) Most of the jobs in fashion need almost the same requirements.
   □   □   3) Assistant designers should have the skills of sketching, draping, drafting and sewing.
   □   □   4) Pattern makers need a working knowledge of computer programs.
   □   □   5) Cutters always create the pattern according to the original design.

Speaking

1. Do you have any expectations for your future? Tell your classmates about your expectations for the future.

2. Discuss with your partner what else you need to do to make your dreams come true.

Language study

1. **Choose the correct expressions in the box to fill in the blanks.**

   | play a big part in | are responsible for | specialize in | in the field of | is short of |

   1) There are many jobs to choose _____ fashion.
   2) He can't get the job because he _____ working experience.
   3) Few people choose to _____ this area because most people don't know computer programs very well.
   4) Insight and intuition always _____ a designer's success.
   5) Assistant designers _____ the sewing of the sample garments, the selection of trimmings and the investigation of the textiles for fabrics.

2. **Complete the sentences with the words or expressions you have learned.**

   1) Zhang Wei and his classmates are talking about their _____.
   2) Constant experimentation with new ideas is a _____ for a fashion designer.
   3) To some degree, designers are the _____ of the fashion companies.
   4) Assistant designers must _____ the designers in _____ way necessary.
   5) The job of cutters requires _____ skill, including _____ with computerized cutting.

3. **Choose the correct sentence in the box for each picture.**

   1) Assistant designers are responsible for the sewing of the sample garments, the selection of trimmings and the investigation of the textiles for fabrics.
   2) Designers are the mainstays of the fashion companies and they must know how to get along with the colleagues.
   3) Cutters always cut the fabrics and other materials into shapes as dictated by the patterns.
   4) Pattern makers need a working knowledge of computer programs.

Unit 11 Career Expectations

Acting it out

1. Tell your group members what you are going to do after graduation and list the reasons for your decision.

2. Invite your group members to give comments or suggestions on your plan.

3. List the improvements you need to make for your future plan.

Your plan	
Reasons	
Suggestions from your classmates	
Improvements	

105

Reading for more

YOUR FIRST JOB

In fashion or in other fields your first job plays an important part in your working life. Here are some tips on how to act in your first job.

1. Be prepared to work hard.

Employers often pay more attention to beginners, so it is proper / likely that you will be observed more than experienced workers.

2. Be observant.

Pay special attention to the way in which other people handle their jobs.

3. Respect your co-workers.

Experienced employees can be a wonderful source of information and advice. Use your co-workers as resources and respect their experience.

4. Find a mentor.

Try to seek out a supervisor who values your abilities and who can give you new assignments, advice and guidance that will help your progress.

5. Become a doer.

A positive and active role will make your job more pleasant. Remember: impress your supervisor with your energy and enthusiasm.

6. Be cooperative.

Help co-workers when you are able to without interrupting your own work schedule. You will then feel freer to ask them for help when you need it.

7. Mistakes will happen.

Learn to acknowledge your errors. No one demands perfection on the job all the time.

Self-check

In this unit I've remembered the following words and expressions:

- ☐ challenge
- ☐ editor
- ☐ consultant
- ☐ wholesaler
- ☐ production

- ☐ employment
- ☐ insight
- ☐ intuition
- ☐ mainstay
- ☐ responsible

Unit 11 Career Expectations

☐ selection
☐ investigation
☐ original
☐ utilization

☐ familiarity

☐ be responsible for

I now understand and can use the following sentences:

_____

_____

Words & Expressions

Words

A

acceptable /əkˈseptəbl/ *adj.* 可接受的；可允许的 (U10)
accessory /əkˈsesərɪ/ *n.* 配饰 (U7)
antique /ænˈtiːk/ *adj.* 古老的，古董的 (U1)
appearance /əˈpɪərəns/ *n.* 外观；容貌 (U7)
armhole /ˈɑːmhəʊl/ *n.* 袖孔 (U3)

B

baste /beɪst/ *v.* 用长针脚疏缝；粗缝 (U6)
bias /ˈbaɪəs/ *n.* 斜边；斜纹 (U3)
blind /blaɪnd/ *adj.* 隐蔽的；视而不见的 (U6)
bodice /ˈbɒdɪs/ *n.* 连衣裙的上部；紧身胸衣 (U3)
bracelet /ˈbreɪslɪt/ *n.* 手镯 (U7)
brocade /brəʊˈkeɪd/ *n.* 锦缎 (U5)
buttonhole /ˈbʌtnhəʊl/ *n.* 纽扣孔 (U3)

108

Words & Expressions

C

candidate /ˈkændɪdɪt/ n. 求职者 (U10)
catwalk /ˈkætwɔːk/ n. (时装表演时模特走的)伸展台 (U1)
challenge /ˈtʃælɪndʒ/ n. 挑战；比赛 (U11)
chiffon /ˈʃɪfɒn/ n. 雪纺绸；薄绸 (U5)
collection /kəˈlekʃən/ n. 时装展销 (U8)
combine /kəmˈbaɪn/ v. 结合，组合 (U3)
commeat /ˈkɒment/ n. 评论 (U1)
complement /ˈkɒmplɪmənt/ v. 搭配；补足 (U7)
conduction /kənˈdʌkt/ v. 传导(热或电) (U5)
construction /kənˈstrʌkʃən/ n. 结构 (U4)
consultant /kənˈsʌltənt/ n. 顾问 (U11)
consumer /kənˈsjuːmə/ n. 消费者 (U9)
convey /kənˈveɪ/ v. 传达；表达 (U8)
coordination /kəʊˌɔːdɪˈneɪʃən/ n. 协调；调节 (U10)
cordial /ˈkɔːdɪəl/ adj. 亲切的 (U9)
cotton /ˈkɒtn/ n. 棉布 (U5)
crush /krʌʃ/ v. 压坏；压伤 (U5)
culmination /ˌkʌlmɪˈneɪʃən/ n. 顶点；巅峰 (U9)

D

dart /dɑːt/ n. 缝褶 (U3)
decoration /ˌdekəˈreɪʃən/ n. 装饰；装饰品 (U2)
decorative /ˈdekərətɪv/ adj. 装饰性的 (U6)
denim /ˈdenɪm/ n. 斜纹粗棉布 (U5)
draft /drɑːft/ v. 起草，草拟 (U3)
drape /dreɪp/ v. 把(织物)披在…上 (U3)
dye /daɪ/ n. 染料；染色 (U5)

E

earring /ˈɪərɪŋ/ n. (常用复数)耳环；耳饰 (U7)
edge /edʒ/ n. 边；边线；边缘 (U6)
editor /ˈedɪtə/ n. 编辑 (U11)
elegant /ˈelɪgənt/ adj. 优雅的；文雅的 (U5)
emcee /ˌemˈsiː/ n. 司仪；节目的主持人 (U8)
emphasize /ˈemfəsaɪz/ v. 突出；强调 (U9)

employer /ɪmˈplɔɪə/ n. 雇主；雇佣者 (U10)
employment /ɪmˈplɔɪmənt/ n. 雇佣 (U11)
even /ˈiːvn/ adj. 均匀的 (U5)
exclusivity /ekˌskluːˈsɪvɪtɪ/ n. 独特性 (U9)
exhibition /ˌeksɪˈbɪʃən/ n. 展览会 (U1)

F

fabric /ˈfæbrɪk/ n. 面料 (U1)
fade /feɪd/ v. 褪色；凋落 (U5)
familiarity /fəˌmɪlɪˈærɪtɪ/ n. 熟悉；通晓 (U11)
fiber /ˈfaɪbə/ n. (动植物的)纤维 (U5)
firm /fɜːm/ adj. 牢固的，稳固的 (U5)
fitting /ˈfɪtɪŋ/ adj. 合身的 (U2)
flat /flæt/ adj. 平坦的 (U4)
flexible /ˈfleksɪbl/ adj. 有弹性的；柔韧的；易弯曲的 (U6)
flow /fləʊ/ v. 垂；飘拂 (U3)
fold /fəʊld/ v. 褶痕；褶缝；褶裥 (U6)
fray /freɪ/ n. 磨损 (U6)
fresh /freʃ/ adj. (指颜色)鲜明的，未褪色的 (U5)
friction /ˈfrɪkʃən/ n. 摩擦 (U5)
frill /frɪl/ v. 起边皱 (U5)

G

gabardine /ˈgæbədiːn/ n. 一种宽松的长袍 (U5)
garment /ˈgɑːmənt/ n. 服装 (U3)
gather /ˈgæðə/ v. 给(衣服)打褶 (U6)
glittering /ˈglɪtərɪŋ/ adj. 闪光的；闪耀的 (U7)
gown /gaʊn/ n. 女长服；礼服 (U5)
graphics /ˈgræfɪks/ n. 图样，图案 (U1)

H

hatch /hætʃ/ v. 策划；孵化 (U2)
hem /hem/ n.(裙子、窗帘等的)下摆，褶边
　　　v. 给…镶边 (U6)
hide /haɪd/ v. 遮挡；把…藏起来 (U6)

Words & Expressions

I

illustration /ˌɪləˈstreɪʃən/ n. (书、杂志等中的)图表，插图 (U1)
indeed /ɪnˈdiːd/ adv. 的确；实在 (U7)
influence /ˈɪnfluəns/ n. 影响；作用 (U1)
initially /ɪˈnɪʃəli/ adv. 开始；最初 (U2)
insert /ɪnˈsɜːt/ v. 插入；嵌入 (U6)
insight /ˈɪnsaɪt/ n. 洞察力 (U11)
inspiration /ˌɪnspəˈreɪʃən/ n. 灵感 (U1)
intermediate /ˌɪntəˈmiːdɪət/ adj. 中间的 (U9)
interview /ˈɪntəvjuː/ n. 面试；面谈
        v. 进行面试 (U10)
interviewer /ˈɪntəvjuːə/ n. 面试官；接见者 (U10)
intuition /ˌɪntjuːˈɪʃən/ n. 直觉 (U11)
investigation /ɪnˌvestɪˈɡeɪʃən/ n. 调查；研究 (U11)
invisible /ɪnˈvɪzəbl/ adj. 看不见的 (U6)
irregulars /ɪˈreɡjuləz/ n. 次品 (U9)

J

jersey /ˈdʒɜːzɪ/ n. 平针织物；紧身套衫 (U3)
jewelry /ˈdʒuːəlrɪ/ n. 珠宝；首饰 (U7)
judgment /ˈdʒʌdʒmənt/ n. 判断 (U2)

K

knit /nɪt/ v. 编织；针织 (U6)

L

layout /ˈleɪaut/ n. 布局；设计 (U2)
leather /ˈleðə/ n. 皮革 (U5)
line /laɪn/ n. 细线；线条 (U1)
linen /ˈlɪnɪn/ n. 亚麻布 (U5)
lining /ˈlaɪnɪŋ/ n. 里料；里子 (U4)
loose /luːs/ adj. 稀松的；未织牢的 (U6)
luxurious /lʌɡˈʒurɪəs/ adj. 十分舒适的，奢侈的 (U5)

M

mainstay /ˈmeɪnsteɪ/ n. 支柱 (U11)

111

maintain /meɪn'teɪn/ v. 保留 (U9)
major /'meɪdʒə/ n. 主修科目
　　　　　v. 主修(某一科目) (U10)
mark /mɑːk/ v. (在…上)做记号 (U3)
measurement /'meʒəmənt/ n. 尺寸；大小 (U3)
mend /mend/ v. 缝补；修补 (U6)
merchandise /'mɜːtʃəndaɪz/ n. 商品；货品
　　　　　v. 销售 (U9)
moisture /'mɔɪstʃə/ n. 湿气；潮湿 (U5)
motion /'məʊʃən/ n. 移动；运动 (U4)
muslin /'mʌzlɪn/ n. 平纹细布 (U3)

N

neat /niːt/ adj. 整齐的；整洁的 (U10)
necklace /'nekləs/ n. 项链 (U7)
needle /'niːdl/ n. 缝衣针 (U6)
nylon /'naɪlɒn/ n. 尼龙 (U5)

O

optimistic /ˌɒptɪ'mɪstɪk/ adj. 乐观的 (U2)
original /ə'rɪdʒənəl/ adj. 原始的；最初的；最早的 (U11)
outfit /'aʊtfɪt/ n. 整套服装 (U7)

P

palette /'pælɪt/ n. 调色板 (U1)
permanent /'pɜːmənənt/ adj. 持久的；永久的 (U6)
phase /feɪz/ n. 阶段；时期 (U9)
photograph /'fəʊtəɡrɑːf/ n. 照片 (U1)
photography /fə'tɒɡrəfɪ/ n. 照相术，摄影 (U1)
polyester /ˌpɒlɪ'estə/ n. 聚酯纤维 (U5)
portfolio /ˌpɔːt'fəʊlɪəʊ/ n. 作品集；代表作 (U10)
position /pə'zɪʃən/ n. 工作职位 (U10)
posture /'pɒstʃə/ n. 姿态 (U8)
preceding /prɪ'siːdɪŋ/ adj. 前面的 (U6)
preference /'prefərəns/ n. 偏爱；钟爱 (U3)
prescribed /priː'skraɪbd/ adj. 规定的；指定的 (U3)

Words & Expressions

present /prɪˈzent/ v. 提供；提交 (U10)
print /prɪnt/ n. 印花布 (U5)
procedure /prəˈsiːdʒə/ n. 步骤；手续 (U3)
process /ˈprəʊses/ n. 过程；进程 (U2)
production /prəˈdʌkʃən/ n. 制造；生产 (U11)
professional /prəˈfeʃənl/ adj. 专业的；内行的 (U8)
prop /prɒp/ n. 道具 (U8)
proportion /prəˈpɔːʃən/ n. 比例；协调 (U2)
psychology /saɪˈkɒlədʒɪ/ n. 心理学 (U9)

R

rehearse /rɪˈhɜːs/ v. 排练 (U8)
responsible /rɪˈspɒnsəbl/ adj. 需负责的 (U11)
résumé /ˈrezjʊmeɪ/ n. 简历 (U10)
retailer /ˈriːteɪlə/ n. 零售商 (U9)
routine /ruːˈtiːn/ n. 表演的成套动作 (U8)

S

sample /ˈsɑːmpl/ n. 样板；样品 (U4)
sandal /ˈsændl/ n. 凉鞋 (U7)
seam /siːm/ n. 线缝；缝口 (U3)
seersucker /ˈsɪəsʌkə/ n. 泡泡纱 (U5)
selection /sɪˈlekʃən/ n. 挑选；选择 (U11)
serviceable /ˈsɜːvɪsəbl/ adj. 能用的；实用的 (U9)
shape /ʃeɪp/ v. 使成形；塑造 (U3)
shed /ʃed/ v. 去掉，除掉 (U5)
shine /ʃaɪn/ v. 出众；超群 (U7)
shoulder /ˈʃəʊldə/ n. 肩；肩膀 (U3)
show-stopper n. (被长时间的掌声所打断的) 精彩表演 (U8)
shrink /ʃrɪŋk/ v. 收缩；萎缩 (U5)
silk /sɪlk/ n. 丝，丝绸 (U5)
sincere /sɪnˈsɪə/ adj. 诚恳的 (U9)
sleeve /sliːv/ n. 袖子 (U3)
slope /sləʊp/ n. 斜面 (U3)
solution /səˈluːʃən/ n. 解决办法 (U4)
source /sɔːs/ n. 资源 (U1)

specialist /ˈspeʃəlɪst/ n. 专业工作者 (U8)
staff /stɑːf/ n. 全体成员 (U4)
stagehand /ˈsteɪdʒhænd/ n. 舞台工作人员 (U8)
stand /stænd/ n. 台；架 (U3)
stitch /stɪtʃ/ n. 线迹；针脚
　　　　　 v. 缝，缝合 (U6)
structure /ˈstrʌktʃə/ n. 结构；构造 (U2)
stuff /stʌf/ n. 资料；东西 (U7)
surface /ˈsɜːfɪs/ n. 表面 (U2)

T

tack /tæk/ v. （通常指在细缝之前用大针脚）粗缝 (U6)
tailored /ˈteɪləd/ adj. (衣服)合身的；专为…而做的 (U3)
target /ˈtɑːɡɪt/ n. 目标 (U9)
technical /ˈteknɪkl/ adj. 技术的；工艺的 (U3)
technology /tekˈnɒlədʒɪ/ n. 技术 (U1)
texture /ˈtekstʃə/ n. 手感；质感；质地 (U10)
theme /θiːm/ n. 主题 (U8)
thread /θred/ n. 线 (U6)
tiny /ˈtaɪnɪ/ adj. 极小的；微小的 (U6)
tiresome /ˈtaɪəsəm/ adj. 令人厌倦的；索然无味的 (U4)
tone /təʊn/ n. 语气；音调 (U9)
translucent /trænzˈluːsnt/ adj. 半透明的 (U2)
trend /trend/ n. 趋势 (U1)
trimming /ˈtrɪmɪŋ/ n. 装饰品，镶边饰物 (U1)
twill /twɪl/ n. 斜纹织物 (U5)

U

uniform /ˈjuːnɪfɔːm/ adj. 统一的 (U5)
up-to-date /ˌʌp tə ˈdeɪt/ adj. 最新的；新近的，新式的 (U10)
utilization /ˌjuːtɪlaɪzeɪʃən/ n. 利用；应用 (U11)

V

venue /ˈvenjuː/ n. 会场；地点 (U8)

W

wholesaler /ˈhəʊlseɪlə/ n. 批发商	(U11)
wool /wʊl/ n. 羊毛；毛料	(U5)
workmanship /ˈwɜːkmənʃɪp/ n. 手艺；工艺	(U9)
wrinkle /ˈrɪŋkl/ n. 皱褶	(U5)

Expressions

A

a sample garment 一件成衣样品	(U4)
at a time 每次	(U6)

B

be responsible for 对…负责任	(U11)
blank paperboard 空白纸板	(U4)
body type 体型	(U7)

C

craft paper 牛皮纸	(U4)

D

depth and breadth of assortments 商品分类的深度和广度	(U9)
dress rehearsal 带妆彩排	(U8)

F

fabric swatch 面料小样	(U1)
fashion comment 时尚评论	(U1)
fashion cycle 流行周期	(U9)
fashion trends 流行趋势	(U9)
finished pattern 完成的样板	(U4)
flat paper pattern 平面纸样	(U4)

115

L

lay out 设计；安排 (U4)

M

merchandising tactics 销售策略 (U9)
model gown 模特样衣 (U9)

P

performing arts 表演艺术 (U8)
price ranges 价格范围 (U9)

R

raw material 原材料 (U1)

S

stage set 舞台背景 (U8)

Text Translations

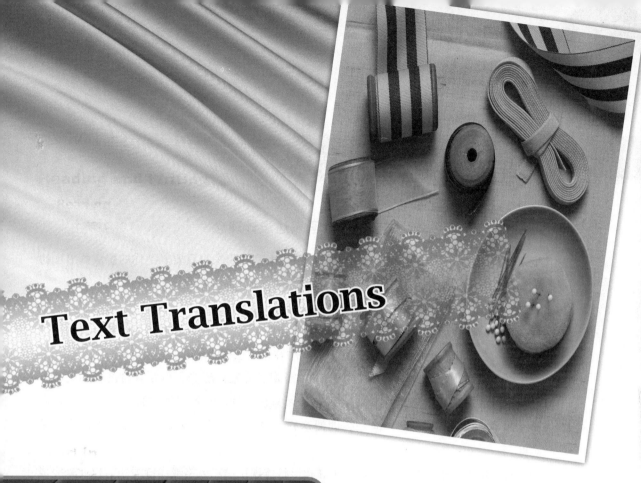

第一单元

设计灵感的来源

对设计师来说,设计灵感几乎可以来自任何地方。书和杂志、展览会、艺术展、国际时事、剧院、音乐、舞蹈、网络、家人照片和世界旅游都是设计师设计灵感的来源。

书和杂志常常有助于获取信息和选取图片风格。行业杂志是流行设计趋势、时装发布信息和时尚评论的主要来源,提供了面料技术新发展的信息。

展览会能启发灵感。在过去的三十年中,举办过印第安艺术展、墨西哥艺术展、埃及艺术展和法国艺术展,它们都对服装设计产生了直接的影响。

感受不同的文化能激发设计灵感。

图片和照片能为设计和图例效果提供丰富的设计灵感来源。

原材料可以成为设计灵感的主要来源。如:面料小样、调色板、辅料、古墙纸、旧时期面料等。

一个优秀的设计师知道如何从各种资源中找出线索,并能消化吸收,最终形成新鲜独特的设计。

117

第二单元

学会通过绘画表达设计理念

作为一名设计师，懂得如何把自己的想法在纸上表达出来很重要。把你的想法画出来的过程不仅能表达设计理念，还能帮助你通过不断的尝试而产生新的灵感。

下面就是一种常见的设计工作方式。

首先，根据设计理念在草图本或者透明纸上画出人体图，画的同时还要不断修改结构，这期间比起人体画得是否正确来说，将注意力集中在设计想法的拓展上更加重要。

为了展示最佳设计理念，在第二阶段，你需要了解流行趋势，也要有自己新的见解。一般说来，形成一个好的理念是要花时间的，因为它本身常常很有挑战性，而且一开始看上去会觉得有些"不对劲儿"，但是，即便这样，也不能放弃。要保持自己的判断力和乐观的态度。

当你对所画作品感到满意时，就用轮廓和形状体现比例、风格或净缝线条，以此来完成整个设计，进而完成设计草图。

最后，不要忘了再加上些外表装饰以及一些细节的添加，并最终完成设计。它们对获得最佳的实验结果同样很重要。

在整个过程中，你都需要考虑选定的面料、装饰及色调。

第三单元

立体裁剪和平面打版

打版是设计过程中重要的一步。打版师可以使用两种方法：立体裁剪和平面打版。

立体裁剪是在人台的样布上进行。打版师在人台上剪裁，这样可以使设计师从立体角度直观地看到服装的比例、线条，必要的话可以随时改变设计。

打版师在人台上工作，必须仔细标注出所有内容：前中线、肩线、接缝、袖笼、扣眼，等等。立体裁剪主要适用于针织物面料或很多又滑又软的质地的面料，也适用于斜纹面料。

平面打版主要根据基本形状或具体测量数据完成制版。平面打版需要在纸上根据预先量好的数据裁剪出服装的不同部位。在准确度测试后，这些版形就成为样板，可以通过移动缝口和缝合线来改变或调整成新的款式。

平面打版主要用于袖子、裤子等在人台上难以实现的部位的裁剪。制作精良的成衣很多都是通过平面打版制作的。

大多数的打版师会把两种方法结合起来使用，比如紧身胸衣、袖子、裤子或裙子等，两种方法均适用。选择哪种方法取决于设计要求、工艺素养以及个人喜好。

第四单元

样衣制作

样衣制作对开发新的服装设计与成衣是非常重要的。虽然样衣制作很乏味，制作过程很枯燥，但是它提供了一个适合模特的完整样板，而且可以无数次地被用来制作大批的成衣。

样衣制作大体要经过纸样裁剪、坯样检验和样衣缝制三个过程。

纸样裁剪是根据服装款式图，由样板师在牛皮纸或空白纸板上画出，并裁剪出样衣的纸样。纸样的大小如同一个普通人一样。

纸样被剪裁好后，由设计助理或样板师铺在白坯布上，按纸样的形状裁剪并进行简略的缝制，这就是坯样检验。这个过程把平面的纸样转化成立体的服装形象，目的在于检测纸样的准确性，一些复杂的结构问题通过这一过程能够得到解决。经过不断的检验和反复修改，最终形成了准确的样板，即成样。

样衣缝制是指依据精确的样板，选用设计师确实需要的面料、里料和辅料缝制成一件成衣。样衣是由样衣工，即工厂里最好的缝纫工完成的。他或她必须了解工厂的缝纫技术，知道如何做出一件完整的成衣。设计团队和样衣工紧密合作，共同来解决结构问题。

样衣不仅要符合人体模型，而且要适合动态的模特，这是非常重要的。设计师可以让模特穿着样衣行走来了解样衣是否舒适和美观。有时还必须制作几件样衣来找到最适合模特的处理方法。

第五单元

多样的服装面料

服装面料是用来制作服装的材料。作为服装基本要素之一，面料在服装风格方面起着很重要的作用。从某种程度上说，面料直接左右着服装的颜色及造型效果。

服装面料可分为：棉、麻、毛、丝绸及人造纤维面料、针织面料及皮革等。不同面料有不同的特性。

棉织品因为轻便、易清洁而多被作为暖季面料来使用。尽管这种面料易缩水、易起皱，但还是常常被用来制作时装、休闲装、内衣及衬衫。

麻织物的导热性和吸湿性好。一般被用来制作休闲装、工作装和夏季服装。

毛织物常常被用作秋冬季节的服装材料。这种材料耐磨、柔软、高雅、保暖性好，常用于制作礼服、西装等高档服装，但这种材料清洁起来比较困难。

丝绸织物质地轻、柔软舒服、色泽鲜艳，但是易褪色。常用于制作高档服装，尤其是女装。

人造纤维面料常用于制作中低价位的服装，虽然耐磨性差，但是质地轻、色彩鲜艳、垂感好，且滑爽舒适，广受欢迎。

皮革面料常被用于制作女士服装或冬季服装，其保暖性好，奢华富贵，但是价格高昂，难以储藏，护理方面要求很高。

面料本身会令人产生设计灵感。有人说过："我的很多服装设计，其灵感就是来源于面料。"

第六单元

手缝

我们都知道机缝和手缝是两种不同的缝制技巧，但对于制作一件好衣服都很重要。

机缝针迹是一种永久针迹，主要用于把服装各个部分缝在一起。手缝用在机缝不太适用的地方，主要是疏缝、装饰针迹及折边缝等。这些在缝纫中都非常重要而且有用。我们来了解一下吧。

1. 初缝针迹是一种基础的手工针迹。把针从面料的背面插进去，一次穿行几针，然后将线拉出面料，如此往复。初缝往往用于将织物缝在一起、修补或试穿等。

2. 回针针迹是手缝针迹中最结实的。针穿过面料，往后回一小段针距，再把针从前面那一针的前方穿出，再进行下一个回针，将针回到前面那一针的顶头穿进。回针针迹主要用于缝纫机难以达到的部位。

3. 暗缝针迹主要用于缝折边及挂面，正面几乎看不见线迹。带针穿过底边的折边，在同一点处从面料中挑出一根线。

4. 之字形针迹主要用于修整面料底边。从左向右缝，在底边抽取细小的针脚，然后穿过服装，针脚保持宽松。这种针迹也可用于在有衬里的服装上做折边针迹。暗之字形针迹可隐藏在衣片和折边间，尤其适用于针织织物，因为这种针迹富有弹性。

5. 暗针针迹用于折边和衣片之间。针穿过底边边缘，挑起面料中的一两根线。织物正面的线迹很小，几乎看不见。

第七单元

时尚饰品

时下，人群中越来越流行佩戴时尚饰品。今年最流行的饰品都有哪些?

1. 珠宝首饰

闪闪发光的珠宝首饰是最为人们所熟知的时尚饰物。首饰那漂亮的颜色，让青少年和儿童也都对它爱不释手。男士经常佩戴手表和戒指，女士则佩戴项链、耳饰和手链等让自己变得更漂亮且富有魅力。

2. 手提包

手提包也是大多数女士和年轻女孩喜爱的流行佩饰，她们需要用手提包把很多物品带到办公室或者学校。除此之外，手提包还可以用来装化妆品，用来保持一天的光鲜亮丽。

3. 鞋和凉鞋

鞋或凉鞋也是备受女士们喜爱的流行饰品之一。多数女士都拥有许多双款式和颜色各异的鞋来搭配不同的服饰。

要想使你普通的外表焕然一新，佩戴饰品绝对是一个不错的好方法。但是，并不是所有的时尚饰品都能很好地衬托你的衣服和身材，搭配不当的话，反而会丑化自己。所以一定要精挑细选你的饰品，学好怎样穿戴饰品，你才能在众人之间脱颖而出。

第八单元

策划一场时装秀

想要推出一台自己的时装秀吗？看完本文之后，你就会觉得这一点也不难！

首先，制定演出方案。

方案中要涵盖时装秀的主题、成衣、模特、佩饰及道具、场地、走道及T台的背景设置以及灯光和音响。如果能考虑到以上这些方面，你就可以对时装秀心中有数了。

第二，和大家讨论你的方案。

如果你对自己的演出方案满意的话，和其他相关人员一起讨论一下，让大家都谈谈自己对这场时装秀的想法。然后确定分工，确保每个人都清楚自己的职责。

第三，落实方案。

模特不仅要姿态美，而且气质要符合你这场秀的主题。找专业人员来帮助，例如表演专业的老师——他们可以帮助模特排练走台路线，还有灯光师、音响师、司仪、舞台工作人员、画外音等。

第四，串演。

要想把演出内容串联起来并且带有专业的味道，你至少需要三遍彩排，包括最后一次带妆彩排。

最后，演出开始之前，检查一切是否准备就绪。

看完本文，你肯定已了解策划一场服装秀所要做的一切了，那么现在你该将计划付诸实践，创造出一台精彩的时装秀了！

第九单元

销售策略

零售商使用的销售策略对维护和吸引顾客是非常有用的。在众多销售策略中，以下的策略要素是最重要的。

高素质的销售人员

与顾客的有效沟通是服装销售的关键。高素质的销售人员不仅要有专业的服装知识，还要懂得顾客心理学，他们知道如何通过得体的穿着、亲切真诚的语气和细致周到的服务赢得顾客的信任。

突出服装流行周期的某个阶段

为了确立自己的形象，每个零售商都首先确定重点推出服装流行周期中的哪个阶段，然后再选定相应的服装商品与该阶段相配。如果一个零售商选择在某些款式正处于导入阶段的时候引进它们，他就能成为时尚的引领者。如果一个零售商更愿意在某些款式的服装处于顶峰期时再引进的话，那他就是时尚潮流的追随者。

质量策略

一般来说，有三种质量水平供零售商来选择，即：①具有顶级面料和工艺制作的顶级水平；②关注质量和工艺，但总是维持一定价格的中等水平；③采用低质面料及工艺，且保持相对低价格的适用水平。一旦商家确立了质量策略，就必须做出更精确的决定，例如：是否不接受任何不是精品的商品，或是否将允许出售次品和二流的商品。

零售商们同样也关注其他的销售策略，例如：价格范围、产品分类的深度和广度、独特性、品牌策略等。无论零售商选择怎样的销售策略，顾客满意永远都是他们的最终目标。

第十单元

作品集

面试准备工作对于每个学生来说都不是件容易事，尤其对学艺术的学生来说更是如此，

他们不仅要准备一份很好的简历，一份合适的作品集对于面试来说也是非常必要的。

大多数艺术类或与设计有关的工作都要求求职者提供一份最有代表性的作品集。面试官或雇主会迅速地看一下你的作品，对你在绘画技巧、设计灵感、对颜色、织物纹理的感觉以及协调性等方面有个初步的印象。因此面试前准备一份优秀的作品集是非常重要的。如下是一些建议：

1. 不必把所有作品都带去面试，8~12幅作品已足够。

2. 一定要保证面试时的作品都确确实实是最新的。因为雇主或面试官很忙，没有足够的时间细看大量作品。事实上他们的兴趣更集中在考察你的作品是否能体现超前一两季的设计理念。

3. 在面试期间，你一定要问面试官是否愿意让你谈一下你的作品或想法，这有助于他们更多地了解你的能力。

4. 作品集一定要保持整洁。

5. 不同的面试要准备不同的作品集，因为适合这个面试的作品集不一定适合另一个面试。

6. 不要把你的作品集留在面试单位，如果要求你把作品留下，你要安排一个取回的时间。

7. 不要抄袭或带去其他艺术家的作品，只展示你自己的作品。

记住这些建议，这对你的求职很有帮助。

第十一单元

服装时尚领域的机遇

时尚领域里充满了各种具有挑战性、令人激动并且高薪的工作机会。在这个领域里，人们可以很轻松就换了工作，哪怕是到一个完全陌生的城市也不必非得换一个全新工种不可。时尚杂志社的编辑、面料设计师、色彩顾问、设计师、批发商、零售商、电视及互联网产品销售、市场顾问、公共关系顾问等均是该领域的职业。在这个领域工作是很有趣的。每种职业都有其自身的要求，而且对于服装时尚的理解在这些行业中显得极为重要。

设计师：最成功的设计师要有良好的教育背景，例如要懂得绘图、立体裁剪、平面裁剪及缝纫，此外洞察力和直觉同样对设计师的成功起着重大的作用。不断尝试新的理念是很有必要的。从某种程度上说，设计师是服装公司的主体，没有他们的创造力，就谈不上系列时装的销售。

助理设计师：助理设计师要负责服装样品缝制、服装配饰选择和面料纺织研究等。此外，他们还要从各个角度协助设计师的工作。

打版师：打版师要根据最初的设计制版，制出的样品最终要用于成衣的制作，因此他们必须在结构、尺寸、生产、裁剪及面料使用等方面技艺娴熟。现在大多数公司为打版师提供电脑，因此打版师需要掌握一定的电脑打版技术。这个职位的薪水很高，因为很少有人选择在这方面发展专长。

裁剪师：裁剪师通常会按照版式的要求将布料或其他材料裁剪成不同形状。这个职业要求具备专业技术能力，还要熟知电脑裁剪技术。

郑重声明

高等教育出版社依法对本书享有专有出版权。任何未经许可的复制、销售行为均违反《中华人民共和国著作权法》，其行为人将承担相应的民事责任和行政责任；构成犯罪的，将被依法追究刑事责任。为了维护市场秩序，保护读者的合法权益，避免读者误用盗版书造成不良后果，我社将配合行政执法部门和司法机关对违法犯罪的单位和个人进行严厉打击。社会各界人士如发现上述侵权行为，希望及时举报，本社将奖励举报有功人员。

反盗版举报电话　　（010）58581999　58582371　58582488
反盗版举报传真　　（010）82086060
反盗版举报邮箱　　dd@hep.com.cn
通信地址　　北京市西城区德外大街4号
　　　　　　高等教育出版社法律事务与版权管理部
邮政编码　　100120

扫描下方二维码或访问链接地址，获取配套资源。